Industrial Chain, Supply Chain and Value Chain in the Energy Industry: Opportunities and Challenges

Industrial Chain, Supply Chain and Value Chain in the Energy Industry: Opportunities and Challenges

Editors

Jiachao Peng
Le Wen
Jianzhong Xiao
Ming Yi
Mingyue (Selena) Sheng

MDPI

Basel • Beijing • Wuhan • Barcelona • Belgrade • Novi Sad • Cluj • Manchester

Editors

Jiachao Peng
Wuhan Institute of Technology
Wuhan
China

Le Wen
The University of Auckland
Auckland
New Zealand

Jianzhong Xiao
China University of Geosciences
Wuhan
China

Ming Yi
China University of Geosciences
Wuhan
China

Mingyue (Selena) Sheng
The University of Auckland
Auckland
New Zealand

Editorial Office
MDPI
St. Alban-Anlage 66
4052 Basel, Switzerland

This is a reprint of articles from the Special Issue published online in the open access journal *Energies* (ISSN 1996-1073) (available at: https://www.mdpi.com/journal/energies/special_issues/supply_and_value_chain_in_energy_industry).

For citation purposes, cite each article independently as indicated on the article page online and as indicated below:

Lastname, A.A.; Lastname, B.B. Article Title. *Journal Name* **Year**, *Volume Number*, Page Range.

ISBN 978-3-7258-1199-1 (Hbk)
ISBN 978-3-7258-1200-4 (PDF)
doi.org/10.3390/books978-3-7258-1200-4

Contents

About the Editors

Jiachao Peng

Dr. Jiachao Peng is a tenured lecturer (the fifth-level of outstanding talent) at the School of Law and Business, Wuhan Institute of Technology. He holds a PhD in economics from the School of Economics and Management at China University of Geosciences (Wuhan). In addition, he is a visiting scholar at Macquarie University in Australia. Dr. Peng is involved in various research projects, including the National Statistical Science Preference Project of the National Bureau of Statistics in China, Excellent Doctoral Dissertation Innovation Fund Project (Central University Funding), Hubei Province Science and Technology Innovation Talents and Services Special Soft Science Research Project, Hubei Province Social Science Fund Project, Hubei Province Education Department Philosophy and Social Science Research Project, and the Wuhan Shuguang Project. Furthermore, he has contributed to several funded research projects, including those under the National Social Science Fund Project, National Natural Science Fund Surface Project, and Ministry of Education Humanities Social Science Research Planning Fund. Dr. Peng has published over 50 papers in SSCI/SCI/CSSCI/CSCD/EI. He has also authored five books. In addition, he is an Editor and Editorial Board of Chinese Sustainable Development Review (Chinese Journal) International of Energy Research (SCI), Complexity (SCI), PLOS ONE (SCI), Social Sciences & Humanities Open, and as a Guest Editor to organize a Special Issue in the Environmental Research (SCI, Top Journal), Environmental Science and Pollution Research (SCI), Energies (SCI), etc. Also, Dr. Peng is an Advisory Board member of Heliyon (SCI). Dr. Peng was approved for the "Chu-Tian Scholar" Program Talent Plan in Hubei Province in November 2023, and the plan includes 5 years of talent project funding.

Le Wen

Dr. Le Wen is a Senior Research Fellow (over the bar) at the University of Auckland's Energy Centre, holding a PhD in Economics. With a focus on both theoretical and applied economics, Dr. Wen's research encompasses energy economics, energy modeling, and advanced data analysis, addressing the critical energy challenges faced by New Zealand's businesses, policymakers, and communities. This work is particularly relevant as New Zealand aims for a net-zero economy, with current research interests including energy efficiency, energy consumption, electricity pricing, the wind–hydro nexus, and green hydrogen. Dr. Wen has played a key role in securing significant grants through collaborative efforts, such as "Reducing Aotearoa's Urban Carbon Emissions—A Critical Pathway to Net-Zero 2050," "Powering NZ's Green-Hydrogen Economy: Next-Generation Electrocatalytic Systems for Energy Production and Storage," and "Sustainable Biomass-Derived Materials to Replace Bitumen for Transport Infrastructure." Dr. Wen's contributions to the field are recognized through publications in leading journals with high impact factors and CiteScores, including *The Energy Journal, Energy Economics, Energy Policy, Journal of Environmental Management, Transportation Research Part D: Transport and Environment,* and *Energy,* covering a broad spectrum of topics from environmental management to sustainable energy development. Dr. Wen is the TRC-Theme Lead (Transport Sustainability and Decarbonisation) for the Transport Research Centre at the University of Auckland and a Special Issue Guest Editor for *Environmental Research, Energies, Frontiers in Environmental Economics,* and *Environmental Science and Pollution Research.* Additionally, Dr. Wen serves as a reviewer for many top-tier journals, including *Energy Economics, Applied Energy, Transport Policy,* and the *Journal of Environmental Management.*

Jianzhong Xiao

Prof. Dr. Jianzhong Xiao is currently the Vice Dean of the School of Economics and Management at China University of Geosciences (Wuhan). He has completed postdoctoral research at the Economic Research Institute of the Chinese Academy of Social Sciences. He is a member of the Hubei Province Foreign Economic Theory Research Association, Executive Director of the Hubei Province Industrial Economic Association, Director of the Hubei Province Economics Association, Director of the Economic Management Research Committee of the China "Double Law" Research Society, Executive Director of the Energy and Resource Systems Engineering Branch of the China Systems Engineering Society, and Executive Director of the China University Energy Economics and Management Innovation Alliance. From March to June 2002, he visited the Center for Enterprise and Economic Development Research (CEEDR) at Middlesex University Business School in the United Kingdom. From August 2007 to August 2009, he was sponsored by the Chinese government to visit Monash University in Australia. In recent years, he has published three books, co-authored three textbooks, and led three projects funded by the National Natural Science Foundation of China, one project funded by the Humanities and Social Sciences Foundation of the Ministry of Education, and one project funded by the later-stage support program of the Ministry of Education. He has also participated in various international cooperations, national social science funds, and national natural science foundation projects. He has published over 60 papers in domestic and international journals. His main research areas are industrial organization theory and energy economics.

Ming Yi

Prof. Dr. Ming Yi is one of the top ten of the thousand outstanding innovation and entrepreneurship mentors in the country, the backbone talent in academia. He holds a PhD in Economics from Huazhong University of Science and Technology, he is a postdoctoral fellow in Theoretical Economics from Tsinghua University, and he is a visiting scholar at The University of Auckland. He mainly teaches courses in Industrial Economics, Special Topics in Industrial Economics, Research Methods and Writing in Economics, Science and Technology Finance, Frontier of Applied Economics, and Special Topics in Chinese Economy for undergraduate and graduate students. He has won the Special Achievement Award and the first prize in Teaching Achievements in Higher Education in Hubei Province (ranking 6th). His main research areas include innovation management and policy, regional sustainable development, and research in the fields of science and technology finance and green finance.

Mingyue (Selena) Sheng

Dr. Mingyue (Selena) Sheng is a Senior Research Fellow in the Energy Centre at the University of Auckland Business School. She is a specialist in the field of applied economics/econometrics and advanced data analysis. Her research is a testament to her profound commitment to environmental sustainability and human behavior, focusing on critical issues such as climate change, carbon emissions reduction, electric vehicle adoption, and the development of sustainable transport infrastructure. Her expertise in empirical economic modeling for transport systems has led to significant contributions in leading academic journals, including *Transportation Research Part A: Policy and Practice*, *Transportation Research Part D: Transport and Environment*, *Transport Reviews*, and *Energy Policy*. Dr. Sheng's work is pivotal in shaping sustainable transport policies and practices, reflecting her dedication to finding innovative solutions for today's environmental challenges.

Preface

In the face of escalating geopolitical conflicts, energy trade wars, and the pressing demand for sustainable development, the global energy sector stands at a crossroads, necessitating a profound transformation of its industrial, supply, and value chains. This Reprint, through its carefully curated collection of ten scholarly papers, embarks on an exploratory journey into the dynamic interplay between digital governance and the energy industry's evolving landscape. With an aim to elucidate the multifaceted challenges and opportunities within the energy sector, this compilation seeks to provide a comprehensive perspective on the role of digitalization in fostering a sustainable, low-carbon future.

The subject matter of this Special Issue is both timely and critical, addressing the urgent need for innovative solutions to expand the energy industry's value chain and improve energy distribution in alignment with market demands. The scope spans across diverse yet interconnected topics, including the impact of China's coal dependency on global carbon emissions, the strategic application of game theory in the oil industry, and the transformative potential of new digital infrastructure (NDI) and industrial robots in enhancing energy efficiency and international competitiveness.

The motivation behind this Special Issue stems from a recognition of the pivotal role that data and digital technologies play in leading the energy transition. Amidst the backdrop of the digital economy's rapid development, this collection aims to shed light on novel governance models, theories, and methods that can navigate the energy sector through its "impossible triangle" of security, economic growth, and efficiency.

This Reprint is intended for a broad audience, encompassing policymakers, industry professionals, academics, and anyone interested in the intersection of energy, sustainability, and digital transformation. The involved authors, hailing from diverse academic backgrounds, bring forth their unique insights and research findings to contribute to a richer understanding of the energy sector's future trajectory.

Acknowledgments are due to all those who have supported the realization of this work. From the researchers who have contributed their groundbreaking studies to the peer reviewers who have ensured the academic rigor of each paper and the editorial team who have seamlessly brought this collection to life, their collective efforts have been instrumental in advancing the discourse on digital governance and sustainability in the energy industry.

This Special Issue invites readers to engage with the complex challenges and innovative opportunities that define the contemporary energy sector. We hope the findings and recommendations presented herein will enrich the academic dialogue and inspire actionable strategies for achieving a more sustainable and efficient global energy system.

We would like to express our sincere gratitude to the editorial department of Energies and Eartha Yang for their trust, patience, and highly professional publishing experience in relation to this special issue. Similarly, we are extremely grateful for the support of the National Natural Science Foundation of China (No. 72303174 and 72273134). These two projects have provided us with a great deal of inspiration to continue our research and exploration of the hot topics in the energy industry supply chain and value chain. We sincerely thank them for their support.

Jiachao Peng, Le Wen, Jianzhong Xiao, Ming Yi, and Mingyue (Selena) Sheng
Editors

energies

Editorial

Industrial Chain, Supply Chain and Value Chain in the Energy Industry: Opportunities and Challenges

Jiachao Peng [1,*], Le Wen [2,*], Jianzhong Xiao [3], Ming Yi [3] and Mingyue Selena Sheng [2]

[1] Law and Business School, Wuhan Institute of Technology, Wuhan 430205, China
[2] Energy Centre, The University of Auckland, Auckland 1010, New Zealand; m.sheng@auckland.ac.nz
[3] School of Economics and Management, China University of Geosciences, Wuhan 430074, China; xjianzhong@cug.edu.cn (J.X.); yiming@cug.edu.cn (M.Y.)
* Correspondence: jiachao.peng@wit.edu.cn (J.P.); l.wen@auckland.ac.nz (L.W.)

Ongoing geopolitical conflicts, frequent energy trade wars, and related issues significantly undermine the globalization of the energy market. The pressing questions of today's and tomorrow's energy transformation revolve around expanding the energy industry's industry chain, supply chain, and value chain, as well as enhancing the market-oriented distribution of energy through innovative and transformative approaches. Currently, data have emerged as a pivotal force driving economic growth, fueling the energy revolution, and propelling the advancement of digital technologies for the creation of a comprehensive global and regional energy market. This shift introduces fresh governance concepts, theories, methods, and models. The traditional energy system's dual challenges of high demand and high emissions exacerbate ongoing coal-power conflicts and impede the market-based reform of oil and gas pipelines. With the rapid digitization of the energy sector and the challenges posed by integrating large-scale renewable energy sources, distributed power supply, and microgrids, there is an urgent need to adopt digital strategies to address the "impossible triangle" of ensuring energy security, economic growth, and efficiency. Thus, exploring digital governance within the energy sector's industry, supply, and value chains is crucial. This exploration aims to enhance the efficiency of market factor allocation within the energy industry amid the digital economy's swift expansion and address the broader issues of energy market reform and global integration.

This Special Issue presents a collection of 10 rigorously researched papers that delve into the opportunities and challenges within the energy industry's industry chain, supply chain, and value chain. Highlighting the pivotal moment facing the energy sector, driven by the rapid transition towards renewable sources [1–3], evolution of the digital economy [3,4], and the pressing demand for sustainable, low-carbon energy solutions [3,5,6], market optimization and integration [7,8], technological innovation and diffusion [6,9], regional and international cooperations [4,5], and environmental governance and regulations [10], this issue contributes to the discourse on navigating the complexities of modern energy systems. It offers insights into leveraging digital transformation for sustainable development, underscoring the integral role of innovative approaches in advancing the global energy transition.

China's reliance on coal-based energy significantly contributes to its carbon emissions, necessitating structural adjustments and accelerated transformation within the coal industry and its associated sectors. A crucial step towards decarbonization involves understanding the CO_2 emission flow from coal production. In this context, Yang et al. provide a foundational analysis of China's coal-based energy sector, identifying key contributors to its carbon emissions and proposing a shift towards distributed renewable energy sources. This study sets the stage for understanding the broader implications of energy production practices and their global impact. Similarly, Sanseverino and Luu expand the discussion to the global transition towards renewable energy technologies, emphasizing

Citation: Peng, J.; Wen, L.; Xiao, J.; Yi, M.; Sheng, M.S. Industrial Chain, Supply Chain and Value Chain in the Energy Industry: Opportunities and Challenges. *Energies* **2024**, *17*, 1467. https://doi.org/10.3390/en17061467

Received: 12 March 2024
Accepted: 15 March 2024
Published: 19 March 2024

the importance of resource management across the energy technology supply chain to achieve sustainability goals.

The role of New Digital Infrastructure in facilitating a sustainable energy transition is thoroughly examined by Fan et al., highlighting its potential to drive green total factor productivity and foster regional cooperation in pollution reduction. This highlights the transformative power of digitalization in the energy sector, offering a new lens through which to view the challenges and opportunities of achieving carbon neutrality. In a study focusing on the digital economy's impact on carbon emissions, Lyu et al. demonstrate the significant potential of digital technologies to enhance energy efficiency and reduce emissions. This research emphasizes the critical role of digitalization in the energy sector's transition towards a more sustainable and low-carbon model. Subsequently, Huang et al. examine the new energy industry's export sophistication and its impact on CO_2 emissions, advocating for policies that enhance the global competitiveness of renewable energy products. Their research underscores the importance of international cooperation and technological innovation in achieving a low-carbon future.

Balhasan et al. explore the application of game theory in optimizing agreements within the oil industry, suggesting innovative approaches to negotiation that can enhance profitability without compromising on environmental standards. This contribution illuminates the complexity of economic interactions in the energy sector and the potential for strategic cooperation to address profitability and sustainability simultaneously. Complementing this, Zheng et al. investigate the effects of market integration on carbon emissions, offering insights into the delicate balance between economic development and environmental protection. Their findings reveal the importance of industrial rationalization and upgrade in mitigating carbon emissions, highlighting the need for targeted policies that support sustainable development.

In a vein of innovation, Zhang et al. delve into the transformative impact of industrial robots on the energy industry, showcasing how technological advancements can optimize production efficiency and contribute to a more sustainable and internationally competitive energy sector. This highlights the intersection of innovation, sustainability, and economic development within the energy industry. Dai et al. investigate the role of new energy vehicles in promoting low-carbon commuting practices within urban settings. Their findings underscore the importance of supporting infrastructure and public awareness to facilitate the widespread adoption of sustainable transportation options, contributing to the broader goal of urban sustainability and carbon emission reduction.

Environmental governance plays a crucial role in the energy transition. Zhang et al. examine the impact of China's Ecological Civilization Pilot Policies (ECPs) on carbon emission reduction within the urban green energy sector, employing a distinct incentive–constraint model to reflect China's unique political landscape. The results show the potential of ECPs in contributing to global carbon emission reduction and sustainability efforts. By navigating the debate between neoclassical economics and the Porter Hypothesis, the study enriches the discourse on environmental regulations and their efficacy in promoting ecological civilization.

This Special Issue presents a comprehensive overview of the current state and future directions of the energy sector, highlighting the critical importance of integrating digital governance, technological innovation, and sustainable practices across the industry, supply, and value chains. By addressing the challenges and harnessing the opportunities presented by the digital economy, geopolitical dynamics, and environmental concerns, the contributions within this issue offer valuable insights and recommendations for policymakers, industry stakeholders, and researchers committed to advancing the global energy transition towards a more sustainable, efficient, and low-carbon future.

Data Availability Statement: No new data were created or analyzed in this study. Data sharing is not applicable to this article.

Acknowledgments: Thanks to all of the authors and peer reviewers for their valuable contributions to the Special Issue 'Industrial Chain, Supply Chain and Value Chain in the Energy Industry: Opportunities and Challenges'.

Conflicts of Interest: The authors declare no conflicts of interest.

References

1. Yang, Z.; Dou, X.; Jiang, Y.; Luo, P.; Ding, Y.; Zhang, B.; Tang, X. Tracking the CO_2 Emissions of China's Coal Production via Global Supply Chains. *Energies* **2022**, *15*, 5934. [CrossRef]
2. Riva Sanseverino, E.; Luu, L. Critical Raw Materials and Supply Chain Disruption in the Energy Transition. *Energies* **2022**, *15*, 5992. [CrossRef]
3. Lyu, K.; Yang, S.; Zheng, K.; Zhang, Y. How Does the Digital Economy Affect Carbon Emission Efficiency? Evidence from Energy Consumption and Industrial Value Chain. *Energies* **2023**, *16*, 761. [CrossRef]
4. Fan, L.; Zhang, Y.; Jin, M.; Ma, Q.; Zhao, J. Does New Digital Infrastructure Promote the Transformation of the Energy Structure? The Perspective of China's Energy Industry Chain. *Energies* **2022**, *15*, 8784. [CrossRef]
5. Huang, K.; Wang, T.; Peng, J.; Sun, L. The Impact of Export Sophistication of the New Energy Industry on Carbon Emissions: An Empirical Study. *Energies* **2023**, *16*, 3846. [CrossRef]
6. Dai, X.; Ma, T.; Zhou, E. New Energy Commuting Optimization under Low-Carbon Orientation: A Case Study of Xi'an Metropolitan Area. *Energies* **2023**, *16*, 7916. [CrossRef]
7. Balhasan, S.; Alnahhal, M.; Shawan, S.; Salah, B.; Saleem, W.; Tabash, M. Optimization of Exploration and Production Sharing Agreements Using the Maxi-Min and Nash Solutions. *Energies* **2022**, *15*, 8970. [CrossRef]
8. Zheng, K.; Deng, H.; Lyu, K.; Yang, S.; Cao, Y. Market Integration, Industrial Structure, and Carbon Emissions: Evidence from China. *Energies* **2022**, *15*, 9371. [CrossRef]
9. Zhang, W.; Peng, J.; Zhang, L. Disruptive Displacement: The Impacts of Industrial Robots on the Energy Industry's International Division of Labor from a Technological Complexity View. *Energies* **2023**, *16*, 3349. [CrossRef]
10. Zhang, P.; Tan, L.; Liu, F. Assessing the Implications of Ecological Civilization Pilots in Urban Green Energy Industry on Carbon Emission Mitigation: Evidence from China. *Energies* **2023**, *16*, 7638. [CrossRef]

energies

MDPI

Article

Tracking the CO_2 Emissions of China's Coal Production via Global Supply Chains

Zheqi Yang [1], Xuming Dou [2], Yuqing Jiang [1], Pengfei Luo [1], Yu Ding [1], Baosheng Zhang [1] and Xu Tang [1,*]

1 School of Economics and Management, China University of Petroleum, Beijing 102249, China
2 Tianjin Branch, CNOOC China Limited, Tianjin 300459, China
* Correspondence: tangxu2001@163.com; Tel.: +86-15611816677

Abstract: Coal's green mining and scientific utilization is the key to achieve the national vision of carbon peak by 2030 and carbon neutrality by 2060. Clarifying the CO_2 flow of coal production is the core part of decarbonization. This study uses an environmental extended multi-regional input–output (EEMRIO) model to analyze the impact of embodied emissions on the indirect CO_2 emission intensity of coal production between China's coal mining sector and 141 countries/regions. It is found that the CO_2 emission intensity of China's coal production was 34.14 gCO_2/MJ in 2014, while the direct and indirect emission intensities were 16.22 gCO_2/MJ and 17.92 gCO_2/MJ, respectively. From 2007 to 2014, the direct emission intensity of China's coal production increased by 23%, while the indirect emission intensity decreased by 30%. The key material and service inputs affecting indirect carbon emissions of coal production in China are electricity service, metal manufacturing, chemical products, coal mining, and transport, which accounted for 85.5% of the total indirect emission intensity of coal production in 2014. Globally, a large portion of CO_2 from Chinese coal production is emitted to meet foreign direct and indirect demands for material and service inputs. Policy implications related to this outcome are further discussed in the study.

Keywords: coal; CO_2 emissions; input–output analysis; China; GTAP

Citation: Yang, Z.; Dou, X.; Jiang, Y.; Luo, P.; Ding, Y.; Zhang, B.; Tang, X. Tracking the CO_2 Emissions of China's Coal Production via Global Supply Chains. *Energies* **2022**, *15*, 5934. https://doi.org/10.3390/en15165934

Academic Editor: Luisa F. Cabeza

Received: 12 July 2022
Accepted: 14 August 2022
Published: 16 August 2022

1. Introduction

On 22 September 2020, Chinese President Xi Jinping addressed, at the General Debate of the General Assembly's seventy-fifth session, that China would scale up its Intended Nationally Determined Contributions (NDCs) by adopting more vigorous action plans and policies and aim to have CO_2 emissions peak before 2030 and achieve carbon neutrality before 2060. After that, the Chinese government developed detailed plans and programs in a series of summits [1]. The goal of carbon neutrality opens the way to deep decarbonization of China's energy system, including accelerating the increase in non-fossil energy development and consumption and reducing coal consumption as the main path to achieve carbon neutrality [2,3].

Due to China's rich coal, poor oil, and less gas energy resource endowment, coal accounts for more than half of China's primary energy sources. Although the scale of coal in China's total energy consumption continues to decline, the short term is still inseparable from coal due to the characteristics of China's resource endowment and the current stage of economic and social development. Xie et al. (2019) reported that in 2025 China's energy consumption demand will be 5.5–5.6 billion tons of standard coal, of which the coal consumption measure is 2.8–2.9 billion tons of standard coal, commanding 50–52% of total energy consumption [4]. Carbon emissions associated with energy production and consumption are an important source of carbon emissions in China, and carbon emissions from coal production and consumption make up 70–80% of China's total carbon emissions [5]. As coal is the largest producer of China's greenhouse gas emissions, the energy conservation and emission reduction in the coal industry will be the most crucial measure for China to respond to global climate change and solve current long-term environmental problems, which promotes the development of a comprehensive understanding of the direct and indirect carbon emissions of coal mining.

1.1. Review of Earlier Works

The research method for direct coal-related carbon emissions is mainly to compute carbon emissions by obtaining activity data of emission sources and the corresponding carbon emission factors. The carbon emission factors are mainly derived from the 2006 IPCC Guidelines for National Greenhouse Gas Inventories released by the United Nations Intergovernmental Panel on Climate Change (IPCC) in April 2006 [6]. There are some scholars who have conducted studies on direct CO_2 emissions from coal-fired power plants using a direct monitoring approach from an engineering perspective [7]. However, none of these studies considered indirect emissions.

Life Cycle Assessment (LCA) is an important environmental management tool, which not only can direct environmental impacts caused by the implementation of the activity being considered but can also analyze the relevant indirect impact. Fossil fuels, mainly coal, release waste substances into the environment as they power the world's economy. Assessing chains of the processes inside the system with environmental analyses such as LCA is crucial [8]. Wang et al. (2018) used the LCA method to evaluate the direct and indirect environmental problems of mining, washing, and transportation in the process of coal mining in China. The indirect emissions mainly involve the impact of carbon emissions caused by fugitive gas in the production process [9]. The research on the boundary of coal-fired power generation systems includes three different stages: coal production, coal transportation, and coal burning. From the perspective of sensitivity analysis, the environmental impact of the coal carbon supply chain can be reduced [10]. Zhou et al. (2020) further refined the coal mining process based on the whole life cycle model, considering the carbon emissions from mining, ventilation, drainage, power consumption, transportation, and post-mining activities. Indirect carbon emissions from post-mining activities were also taken into account [11]. Burchart-Korol et al. (2016) developed an environmental LCA model applied to coal mining operations, which not only applies to greenhouse gas (GHG) emission assessment but is also connected with the ReCiPe system to identify damage categories such as human health, ecosystem, and resources [12].

Input–output (IO) analysis is commonly used to quantify embodied energy [13], embodied CO_2 [14], and embodied $PM_{2.5}$ emissions [15]. There are many databases that provide IO tables of embodied energy, such as GTAB and EXIOBASE [16]. Compared with the EXIOBASE database, the input–output model of GTAB includes more inter-country trade and is more suitable for studying national emissions. All goods and services produced by an economy are directly or indirectly linked to energy use and, depending on the type of fuel used, to carbon dioxide emissions [17]. Davis and Caldeira (2010) calculated carbon emissions at the global sectoral scale using an EEMRIO [18]. Zhou et al. (2010) combined an IO table with the energy consumption data by sector to estimate embodied carbon emissions in the international trade of China in 2007 [19]. Daly et al. (2015) estimated upstream CO_2 emissions from current and future energy technologies in the UK using a multi-regional environment extended input–output (EEMRIO) model, and explicitly simulated direct and indirect CO_2 emissions from energy supply and infrastructure technologies within the national ESOM (TIMES model) [20]. Some studies account for sector-specific direct and indirect carbon emissions based on sectoral emission intensity and intersectoral economic linkages. Pan et al. (2020) used an IO model to account for the sectoral-scale CO_2 emissions of China, including the oil and gas sector [21]. In addition, a system considering material flow to analyze the embodied carbon emissions of aluminum-containing commodities in China's international trade from 2008 to 2017 has also been developed [22].

1.2. Aim of This Study

Green mining and the scientific utilization of coal are key to achieving the national vision of carbon peaking by 2030 and carbon neutrality by 2060. Coal-based energy structure is the main source of carbon emissions in China, which requires the coal industry and other industries closely connected with the coal mining sector to adjust their structure and accelerate transformation. Therefore, this paper selects the coal mining sector as the research subject, calculates the CO_2 emission intensity of the sector based on the EEMRIO approach, and explores the sustainable development model of the coal industry. This paper is organized as follows: Section 2 explains the method and data, Section 3 describes the results and discussion, and Section 4 presents conclusions and policy implications.

2. Materials and Methods

2.1. Methodology for Accounting CO_2 Emission Intensity

The input–output (IO) model was proposed by Leontief in the 1930s [23], which is mainly through the formulation of the IO table and establishes the corresponding mathematical model to reflect the national economic system of interdependence and the restriction relationship between different departments. Multi-regional input–output models are gradually used to quantitatively analyze the environmental impacts of trade activities between countries or regions, including $PM_{2.5}$, CO_2 [15,17]. This method is used to analyze the direct and indirect CO_2 emissions of coal production in major coal-producing countries (Figure 1). The basic equation is shown in Equation (1):

$$C = E(I-A)^{-1}M = ELM \quad (1)$$

where there are 141 countries or regions and each region has 57 sectors, and C is an 8037×8037 vector representing the complete CO_2 emissions. $E = \frac{e}{x'}$, which is the CO_2 direct emission coefficients of economic sectors; L is an 8037×8037 Leontief inverse, which is also called complete emission factor matrix; M is an 8037×8037 matrix of intermediate demand. The Equation (2) is expressed as a matrix:

$$\begin{pmatrix} C_{1\,1} & C_{1\,2} & \cdots & C_{1\,8037} \\ C_{2\,1} & C_{2\,2} & \cdots & C_{2\,8037} \\ \vdots & \vdots & \ddots & \vdots \\ C_{8037\,1} & C_{8037\,2} & \cdots & C_{8037\,8037} \end{pmatrix} = \begin{pmatrix} E_1 & 0 & \cdots & 0 \\ 0 & E_2 & \cdots & 0 \\ \vdots & \vdots & \ddots & \vdots \\ 0 & 0 & \cdots & E_{8037} \end{pmatrix} \begin{pmatrix} L_{1\,1} & L_{1\,2} & \cdots & L_{1\,8037} \\ L_{2\,1} & L_{2\,2} & \cdots & L_{2\,8037} \\ \vdots & \vdots & \ddots & \vdots \\ L_{8037\,1} & L_{8037\,2} & \cdots & L_{8037\,8037} \end{pmatrix} \begin{pmatrix} M_{1\,1} & M_{1\,2} & \cdots & M_{1\,8037} \\ M_{2\,1} & M_{2\,2} & \cdots & M_{2\,8037} \\ \vdots & \vdots & \ddots & \vdots \\ M_{8037\,1} & M_{8037\,2} & \cdots & M_{8037\,8037} \end{pmatrix} \quad (2)$$

Indirect emissions from production in the coal sector in each country are summed in the corresponding columns of the C matrix in Equation (2). Take China as an example: $\sum_{i=1}^{8037} C_{i\,186}$ is indirect emissions from China's coal production. The MRIO model endogenously calculates not only the domestic output, but also the output in all other regions resulting from intermediate products, which is embodied in international trade. The summation by sector and country can be used to analyze the embodied emissions from coal production of different sectors and countries.

Indirect emission intensity from China's coal production is shown in Equation (3):

$$I_{ind,CHN} = \frac{\sum_{i=1}^{8037} C_{i\,186}}{Q_{coal,CHN}} \quad (3)$$

where $I_{ind,CHN}$ is the indirect emission intensity of China's coal production, $\sum_{i=1}^{8037} C_{i\,186}$ is the indirect emissions from coal production in China, and $Q_{coal,CHN}$ represents the annual coal production in China.

Direct emission intensity from China's coal production is shown in Equation (4):

$$I_{d,CHN} = \frac{C_p}{Q_{coal,\ CHN}} \quad (4)$$

where $I_{d,CHN}$ is the direct emission intensity of China's coal production and C_p is the production-based CO_2 emissions in China.

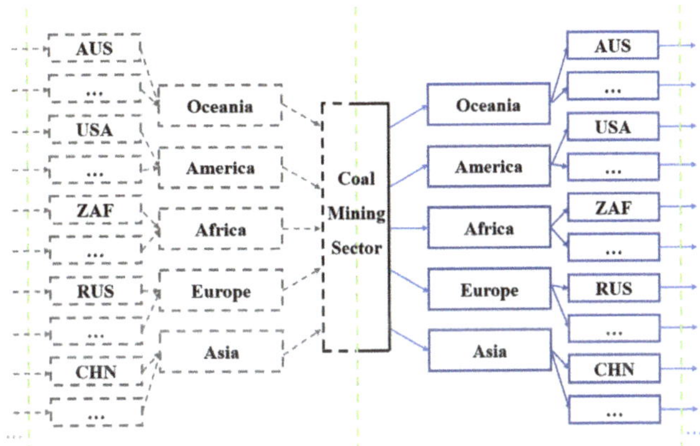

Figure 1. Carbon accounting framework for the coal mining sector.

2.2. Data

In this study, the global production-based CO_2 emissions data and world IO tables were obtained from the latest GTAP 10 [24], which is commonly used in health and environmental research [25], for example, $PM_{2.5}$ and CO_2 accounting studies [26,27]. The Global Trade Analysis Project (GTAP) database provides the world economy for 4 reference years (2004, 2007, 2011, and 2014) and distinguishes 65 sectors, up from 57 in the previous release, in each of the 141 countries/regions. For each country/region, the database presents values of production and intermediate and final consumption of materials and services in millions of US dollars. We mainly analyze intermediate consumption data for the coal sector in 11 countries, as shown in Figure 2. Annual data on coal production by country come from the IEA [28].

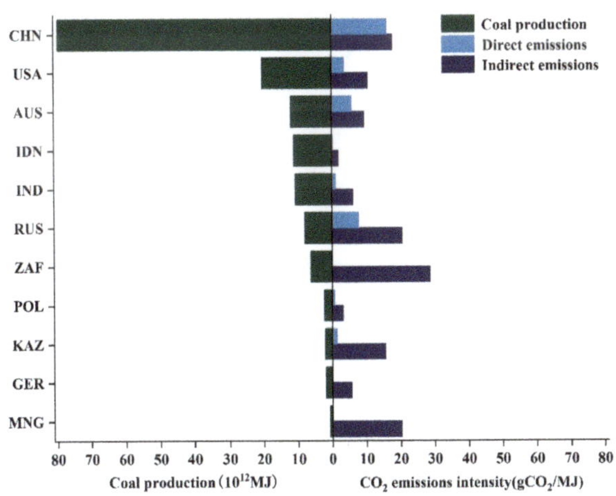

Figure 2. CO_2 emission intensity of coal production in different countries.

3. Results and Discussion

3.1. CO_2 Emission Intensity of Coal Production in Major Coal-Producing Countries

According to the model and data advantage of the input–output approach, we can obtain the CO_2 emission intensity of the world's major coal production in 2014, as shown in Figure 2. The CO_2 emission intensity of coal production in different countries is between 2.21 gCO_2/MJ and 34.14 gCO_2/MJ. China, South Africa, and Russia have large coal production emission intensities, which are 34.14 gCO_2/MJ, 28.61 gCO_2 /MJ, and 28.41 gCO_2/MJ, respectively. At the same time, the CO_2 emission sources from coal production vary widely in different countries. The direct and indirect emission intensity of coal production can be distinguished by the input–output method, as shown in Figure 1. Direct emission intensity refers to the CO_2 emissions of different resource productions, while indirect emission is associated with the material and service inputs in the production process.

The direct CO_2 emission intensity of coal production in 11 coal-producing countries ranges from 0.01 gCO_2/MJ to 16.22 gCO_2/MJ, which is related to the differences in coal mining exploitation in each country. For instance, although China is rich in coal resources, its resource endowment and long-term strong demand have led to the increasing depth of coal mining [29]. Deep coal mine development activities are an important reason for China's direct CO_2 emission intensity ranking first in 2014. Russia's underground coal resources account for 37% of the total resources [30]. Under the circumstance of increasing mining difficulty, the direct carbon emission intensity of coal production was 7.89 gCO_2/MJ in 2014. However, the direct CO_2 emission intensity of coal mining in South Africa is only 0.02 gCO_2/MJ, which has rich open-pit coal resources and superior mining conditions. The indirect CO_2 emission intensity of coal production in 11 coal-producing countries ranges from 1.89 gCO_2/MJ to 28.59 gCO_2/MJ, which is mainly associated with the material and service inputs of the coal mining sector. In addition, the differences in industrial structure, trade structure, and energy structure between countries play more important roles in indirect emissions. Indirect CO_2 emissions from coal production are high in South Africa, Russia, and China, with carbon emission intensities of 28.59 gCO_2 /MJ, 20.52 gCO_2 /MJ, and 17.92 gCO_2 /MJ, respectively.

Figure 3 illustrates the distribution of direct and indirect CO_2 emission intensity in major coal-producing countries in 2014. South Africa, Mongolia, and Kazakhstan are coal producers with high indirect emission intensities. China and Russia are coal producers, which both have high direct and indirect emissions intensities. In general, indirect emissions are higher than direct emissions in most coal-producing countries. In 2014, indirect CO_2 emission intensity accounted for more than 80% of total CO_2 emission intensity in each coal-producing country. Therefore, indirect carbon emission intensity has a significant effect on the overall carbon intensity of coal production. The next section analyzes the differences in indirect emission intensity of coal production between countries from the import/export trade.

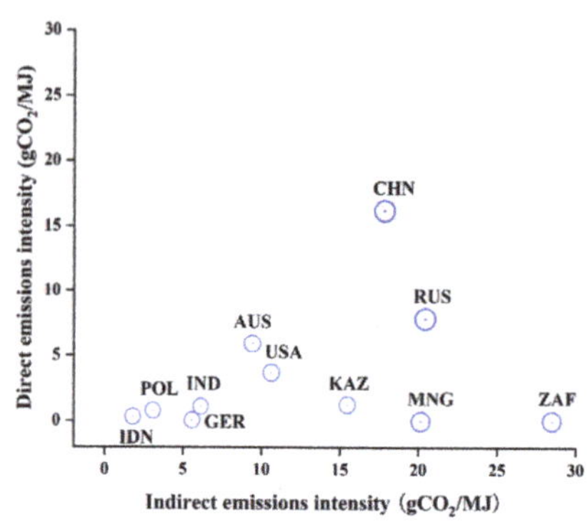

Figure 3. Coal-producing countries' CO_2 emission intensity distribution.

3.2. Indirect CO_2 Emission Intensity of Coal Production from Material and Service Inputs

Compared to the bottom-up approach, the top-down input–output method can also find the key factors affecting the CO_2 emission intensity of coal production. Table 1 shows the 20 major material and service inputs, which accounted for 97.56% of the total indirect CO_2 emission intensity of China's coal production in 2014. Electricity ranks first with 9.18 gCO_2/MJ indirect emission intensity, accounting for half of the total indirect emission intensity. Above all, China's power mix is dominated by coal, which accounted for more than 75% of the total power generation in 2014 [28], resulting in high overall emissions from the power sector. Secondly, the depth of coal mining in China has been increasing as shallow coal resources are depleted [31], leading to an increment in electricity consumption as coal mining becomes harder. The indirect emission intensities caused by ferrous metals, machinery and equipment, chemical products, and metal products are also relatively high, which are 2.31 gCO_2/MJ, 1.07 gCO_2/MJ, 0.75 gCO_2/MJ, and 0.73 gCO_2/MJ, respectively. It shows that there is a great demand for steel and chemicals. In addition, it can also be found in Table 1 that China's coal production leads to the indirect emission intensity of the coal mining sector reaching 0.65 gCO_2/MJ, which further reflects China's current coal-based energy structure.

Table 1. Distribution of indirect CO_2 emission intensity of China's coal production in 2014.

Ranking	Material and Service Input	Indirect CO_2 Emission Intensity
1	Electricity	9.18
2	Ferrous metal	2.31
3	Machinery and equipment	1.07
4	Chemical products	0.75
5	Metal products	0.73
6	Coal	0.65
7	Transport	0.62
8	Wood products	0.40
9	Business services	0.35
10	Petroleum, coal products	0.35
11	Financial services	0.31
12	Mineral products	0.27
13	Trade	0.17
14	Construction	0.08
15	Sea transport	0.08
16	Electrical and electronic equipment	0.06
17	Water	0.04
18	Manufactures	0.03
19	Air transport	0.02
20	Other extraction (mineral)	0.02

In this study, 20 major inputs of material and service are combined into 7 categories of material and service inputs to further analyze the temporal variation trend of indirect CO_2 from coal mining in China and the differences in indirect emission intensity among major coal-producing countries. Electricity service includes electricity; transportation services comprise transport, air transport, and sea transport; extraction services involve coal, water, and other extractions (mineral); metal manufacturing covers ferrous metal, metal products, machinery and equipment, and electrical and electronic equipment; other manufacturing includes wood products, mineral products, and manufactures; support services incorporate business services, trade, financial services, and construction; and refining and chemicals comprise petroleum and coal products and chemical products.

Comparing the CO_2 emission intensity of China's coal production in 2007 and 2014, it can be found that the emission intensity of coal production in 2007 was 38.75 gCO_2/MJ, of which the direct emission intensity was 13.20 gCO_2/MJ, and the indirect emission intensity was 25.55 gCO_2/MJ. In 2014, the emission intensity of coal production was 34.14 gCO_2/MJ, of which the direct emission intensity was 16.22 gCO_2/MJ and the indirect emission intensity was 17.92 gCO_2/MJ. With the depletion of shallow coal resources, the depth of coal mining in China has been increasing, and the direct emission intensity has increased by 23%. Figure 4 shows that the indirect emission intensity of coal production in China decreased by 30% from 2007 to 2014, and embodied emission intensity of electricity service, transportation services, extraction services, metal manufacturing, and other manufacturing decreased by 26%, 62%, 41%, 30%, and 35%, respectively. The decline in the embodied emission intensity of the real economy sector is potentially due to technological progress, energy efficiency improvement, and the adjustment of China's energy structure. The end consumption of coal decreased from 43% in 2007 to 39% in 2014, among which the share of coal power in China's power structure dropped from 81% in 2007 to 73% in 2014 [28]. For support services, including financial services, business services, and trade, the proportion of embodied CO_2 emission intensity of coal production increased by 31%, reflecting the increasing vitality of China's financial market and commercial services as well as the increasing participation of financing activities in the coal sector [32].

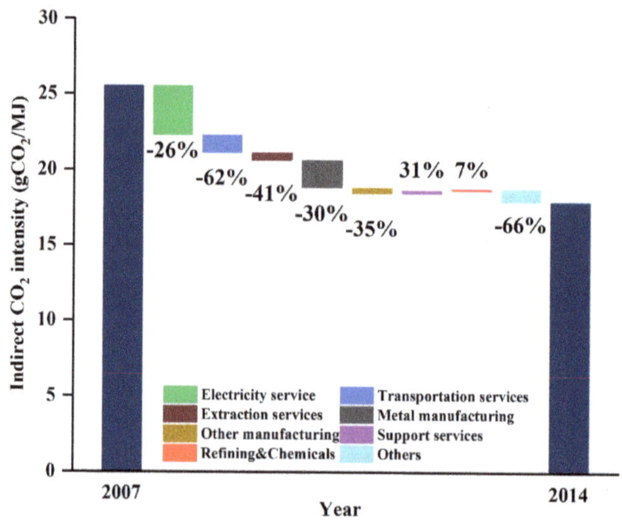

Figure 4. Changes in the indirect emission intensity of coal production in China.

Figure 5 portrays that the structure of the embodied carbon emission intensity from coal production is significantly different between China and the US. China's electricity service accounts for 51% of embodied emission intensity, compared with 31% for the United States. Due to China's coal resource endowment and long-term strong demand, coal mining depth continues to increase, resulting in increased electricity consumption in production activities. Meanwhile, there are differences in the power generation structure between China and the United States. In 2014, 73% of China's electricity came from coal, compared to 39% in the United States [28]. China's embodied emission intensity from extraction services was 4%, while that from USA was less than 1%, which further reflects China's coal-based energy structure. Transportation services is another difference between China and the United States in the structure of embodied emission intensity from coal production. The United States accounted for 35% of embodied emission intensity in transportation serves, while China only accounted for 4%. This is mainly due to China's developed public transportation system and the large number of coal power plants in China, which facilitate local consumption of coal production.

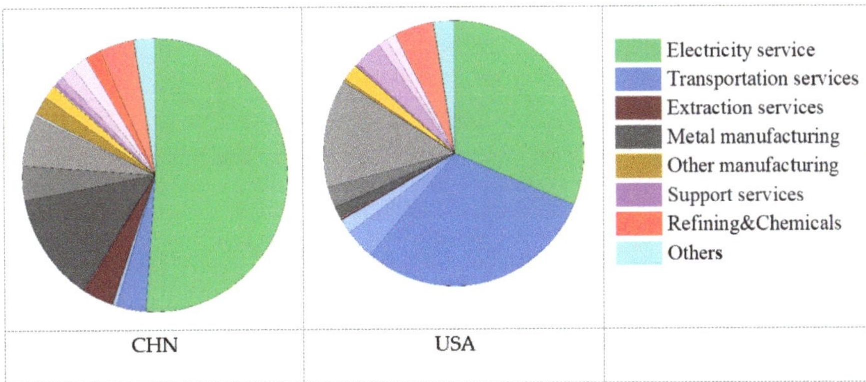

Figure 5. Embodied emission intensity structure of coal production.

3.3. Coal Production CO_2 Emissions Embodied in Trade

In this section, the representative coal-producing countries in each continent are selected to analyze the embodied CO_2 emissions caused by the participation of the coal mining sector in international trade. In 2014, the embodied emissions from international trade related to China's coal mining sector were mainly in Asia, accounting for 64.1%, and are closely related to trade with Japan, South Korea, Thailand, and India, contributing 0.8% of China's total indirect emissions from coal production. Outside Asia, countries with high trade links to China's coal mining sector include the US, Russia, and Australia. However, the indirect carbon emissions of China's coal contributed by international trade only accounts for 1.13%, and most of the embodied emissions of material and service inputs are in China.

As shown in Figure 6, the embodied emissions from international trade related to Russia's coal mining sector are mainly in Asia and Europe, accounting for 42% and 41.7%, respectively. The international trade emissions associated with South Africa's coal mining sector are mainly in Asia, accounting for 66.3%. The embodied emissions from international trade associated with the US's coal mining sector are mainly in Europe and Asia, accounting for 53.2% and 41.7%, respectively. Asia accounts for 76% of international-trade-related emissions from Australia's coal mining sector. In the coal mining sectors of Russia, South Africa, the US, and Australia, China is the largest importer of trade, accounting for 1.24%, 1.28%, 4.34%, and 3.58% of their total indirect emissions from coal production, respectively. Therefore, coal production in Russia, South Africa, the United States, and Australia contributed 0.2 Mt, 0.23 Mt, 0.94 Mt, and 0.41 Mt of CO_2 emissions in China in 2014.

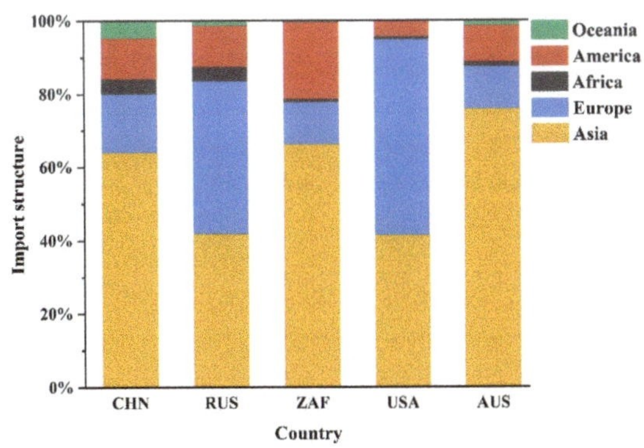

Figure 6. Trade-related CO_2 emission structure from coal production.

4. Conclusions and Policy Implications

(1) In 2014, the CO_2 emission intensity of China's coal production was 34.14 gCO_2/MJ, of which the direct and indirect emission intensities were 16.22 gCO_2/MJ and 17.92 gCO_2/MJ, respectively. From 2007 to 2014, the direct emission intensity of China's coal production increased by 23%, while the indirect emission intensity decreased by 30%. Compared with other coal-producing countries, China has high direct and indirect emission intensity in coal production mining.

(2) The key material and service inputs affecting indirect carbon emissions of coal production in China are electricity, ferrous metal, machinery and equipment, chemical products, metal products, coal mining, and transport, which accounted for 85.5% of the total indirect emission intensity of coal production in 2014. It is worth noting that China's coal mining sector contributes 4% of indirect emissions to coal production, which is much higher than other coal-producing countries.

(3) China's coal production sector is mainly traded with Japan, South Korea, Thailand, and India. All import trade accounts for 0.8% of the total indirect emissions from coal production in China. However, China is the largest import source of material and service inputs for coal production in South Africa, the United States, Russia, Australia, and other coal-producing countries, accounting for 1.24%, 1.28%, 4.34%, and 3.58% of their total indirect emissions from coal production, respectively.

Based on the conclusion of this study, policy recommendations are given for the reduction in CO_2 direct emissions from coal production and CO_2 embodied emissions from trade, respectively. First, on the production side, China's mining difficulty aggravated by resource exhaustion actively promote the research, development, and application of carbon-negative technology represented by carbon capture, utilization, and storage (CCUS) in coal mining. This can alleviate CO_2 emissions in the coal supply chain and industrial chain from the source. Secondly, more than half of the indirect emissions of China's coal production come from electricity service. In 2007 and 2014, the proportion of coal power in China reached 81% and 73%, respectively. However, the proportion of coal power in China bucked the trend and rose to more than 70% during 2021. While China is aggressively pursuing carbon neutrality, its coal-based electricity mix is unlikely to change radically anytime soon. Because of China's vast territory, wind, light, biomass, and other resources are rich. By enhancing the complementary supply of distributed renewable energy electricity including wind power, solar, and biomass in coal production areas in accordance with local conditions, the embodied emissions of the electricity service input in coal mining can be reduced and energy structure adjustment can also be promoted.

Author Contributions: Conceptualization, X.T. and B.Z.; formal analysis, Y.D.; resources, X.D.; data curation, P.L.; writing—original draft preparation, Z.Y.; writing—review and editing, Y.J. All authors have read and agreed to the published version of the manuscript.

Funding: This research was funded by the National Natural Science Foundation of China (71673297, 72174206), and thanks are given to the Science Foundation of the China University of Petroleum, Beijing (ZX20200109), for sponsoring this research.

Institutional Review Board Statement: Not applicable.

Informed Consent Statement: Not applicable.

Data Availability Statement: Not applicable.

Conflicts of Interest: The authors declare that they have no known competing financial interest or personal relationship that could have appeared to influence the work reported in this paper.

References

1. Xinhuanet. 2021. Available online: http://www.xinhuanet.com/world/cop15/index.htm (accessed on 12 October 2021).
2. He, G.; Lin, J.; Sifuentes, F.; Liu, X.; Abhyankar, N.; Phadke, A. Rapid cost decrease of renewables and storage accelerates the decarbonization of China's power system. *Nat. Commun.* **2020**, *11*, 3780. [CrossRef] [PubMed]

3. Fu, C.; Yu, H.C.; Bian, Z.F.; Ying, D.Y. How to handles the crisis of coal industry in China under the vision of carbon neutrality. *J. China Coal Soc.* **2021**, *46*, 1808–1820. [CrossRef]
4. Xie, H.P.; Wu, L.X.; Zheng, D.Z. Prediction on the energy consumption and coal demand of China in 2025. *J. China Coal Soc.* **2019**, *44*, 1949–1960. [CrossRef]
5. Meng, L.; Li, Y.; Zhang, S. Regional differences of carbon emissions in the coal industry based on FPGA and wireless sensors. *Microprocess. Microsyst.* **2020**, 103489. [CrossRef]
6. Eggleston, H.S.; Buendia, L.; Miwa, K.; Ngara, T.; Tanabe, K. *2006 IPCC Guidelines for National Greenhouse Gas Inventories*; Institute for Global Environmental Strategies (IGES): Tokyo, Japan, 2006; ISBN 4-88788-032-4.
7. Gao, T.; Jin, P.; Song, D.; Chen, B. Tracking the carbon footprint of China's coal-fired power system. *Resour. Conserv. Recycl.* **2022**, *177*, 105964. [CrossRef]
8. Czarnowska, L.; Stanek, W.; Pikon, K.; Nadziakiewicz, J. Environmental quality evaluation of hard coal using LCA and exergo-ecological cost methodology. *Chem. Eng. Trans.* **2014**, *42*, 139–144.
9. Wang, J.; Wang, R.; Zhu, Y.; Li, J. Life cycle assessment and environmental cost accounting of coal-fired power generation in China. *Energy Policy* **2018**, *115*, 374–384. [CrossRef]
10. Wang, C.; Mu, D. An LCA study of an electricity coal supply chain. *J. Ind. Eng. Manag.* **2014**, *7*, 311–335. [CrossRef]
11. Zhou, A.; Hu, J.; Wang, K. Carbon emission assessment and control measures for coal mining in China. *Environ. Earth Sci.* **2020**, *79*, 461. [CrossRef]
12. Burchart-Korol, D.; Fugiel, A.; Czaplicka-Kolarz, K.; Turek, M. Model of environmental life cycle assessment for coal mining operations. *Sci. Total Environ.* **2016**, *562*, 61–72. [CrossRef]
13. Feng, C.; Xu, T.; Jin, Y.; Wang, X.; Zhang, B. Inter-industry path decomposition of embodied oil export changes in China. *Pet. Sci. Bull.* **2017**, *4*, 546–556. [CrossRef]
14. Tang, X.; Jin, Y.; Wang, X.; Wang, J.; McLellan, B.C. Will China's trade restructuring reduce CO_2 emissions embodied in international exports? *J. Clean. Prod.* **2017**, *161*, 1094–1103. [CrossRef]
15. Jiang, Y.; Tang, X.; Zhao, X.; Höök, M. Mitigation strategies of air pollution: Case studies of China and the United States from a consumption perspective. *Environ. Sci. Policy* **2022**, *128*, 24–34. [CrossRef]
16. Wiedmann, T. A review of recent multi-region input–output models used for consumption-based emission and resource accounting. *Ecol. Econ.* **2009**, *69*, 211–222. [CrossRef]
17. Machado, G.V. Energy use, CO_2 emissions and foreign trade: An IO approach applied to the Brazilian case. In Proceedings of the Thirteenth International Conference on Input–Output Techniques, Macerata, Italy, 21–25 August 2000.
18. Davis, S.J.; Caldeira, K. Consumption-based accounting of CO_2 emissions. *Proc. Natl. Acad. Sci. USA* **2010**, *107*, 5687–5692. [CrossRef]
19. Zhou, H.; Yang, Y. Carbon emissions embodied in the international trade of China. *Adv. Mater. Res.* **2010**, *171–172*, 183–187. [CrossRef]
20. Daly, H.E.; Scott, K.; Strachan, N.; Barrett, J. Indirect CO_2 emission implications of energy system pathways: Linking IO and TIMES models for the UK. *Environ. Sci. Technol.* **2015**, *49*, 10701–10709. [CrossRef]
21. Pan, L.; Zhang, T.; Li, W.; Li, Z.; Zhou, C. Sector-level evaluation of China's CO_2 emissions: Trend evolution and index ranking. *J. Clean. Prod.* **2010**, *286*, 125453. [CrossRef]
22. Li, Q.; Duan, H.; Li, T.; Zhou, Y.; Chen, Y.; Zhong, R. Embodied carbon emissions of aluminum-containing commodities in international trade: China's perspective. *Clim. Change* **2021**, *166*, 47. [CrossRef]
23. Leontief, W.W. The structure of American economy, 1919–1939: An empirical application of equilibrium analysis (No. HC106. 3 L3945 1951). *Econom. Soc.* **1951**, *19*, 351–353. [CrossRef]
24. Aguiar, A.; Chepeliev, M.; Corong, E.L.; McDougall, R.; van der Mensbrugghe, D. The GTAP Data Base: Version 10. *J. Glob. Econ. Anal.* **2019**, *4*, 1–27. [CrossRef]
25. Ou, J.; Huang, Z.; Klimont, Z.; Jia, G.; Zhang, S.; Li, C.; Meng, J.; Mi, Z.; Zheng, H.; Shan, Y.; et al. Role of export industries on ozone pollution and its precursors in China. *Nat. Commun.* **2020**, *11*, 5492. [CrossRef] [PubMed]
26. Meng, J.; Liu, J.; Xu, Y.; Guan, D.; Liu, Z.; Huang, Y.; Tao, S. Globalization and pollution: Tele-connecting local primary PM2.5 emissions to global consumption. *Proc. R. Soc. A Math. Phys. Eng. Sci.* **2016**, *472*, 20160380. [CrossRef] [PubMed]
27. Mi, Z.; Meng, J.; Guan, D.; Shan, Y.; Song, M.; Wei, Y.M.; Liu, Z.; Hubacek, K. Chinese CO_2 emission flows have reversed since the global financial crisis. *Nat. Commun.* **2017**, *8*, 1712. [CrossRef] [PubMed]
28. IEA. 2021. Available online: https://www.iea.org/fuels-and-technologies/coal (accessed on 30 November 2021).
29. Liang, Y. Strategic thinking of simultaneous exploitation of coal and gas in deep mining. *J. China Coal Soc.* **2016**, *41*, 1–6. [CrossRef]
30. Kuznetsov, P.N.; Ilyushechkin, A.Y. Coal resources, production and use in the Russian Federation. In *The Coal Handbook: Towards Cleaner Production*; Woodhead Publishing: Sawston, UK, 2013; pp. 148–168. [CrossRef]
31. Xie, H.P.; Gao, F.; Ju, Y.; Gao, M.Z.; Xie, L.Z. Quantitative definition and investigation of deep mining. *Mtan Xuebao/J. China Coal Soc.* **2015**, *40*, 1–10. [CrossRef]
32. Yang, L.; Xu, M.; Fan, J.; Liang, X.; Zhang, X.; Lv, H.; Wang, D. Financing coal-fired power plant to demonstrate CCS (carbon capture and storage) through an innovative policy incentive in China. *Energy Policy* **2021**, *158*, 112562. [CrossRef]

energies

Editorial

Critical Raw Materials and Supply Chain Disruption in the Energy Transition

Eleonora Riva Sanseverino * and Le Quyen Luu

Department of Engineering, University of Palermo, Viale delle Scienze Bld. 9, 90128 Palermo, Italy
* Correspondence: eleonora.rivasanseverino@unipa.it

The energy transition towards efficient energy production, transport, and use, renewable energy (RE) technologies and innovative energy management brings benefits to reducing greenhouse gas (GHG) emissions and achieving climate targets. The transition requires resources, minerals, metals, and materials for RE technologies themselves, for example, solar photovoltaics (PV), hydrogen fuel cell vehicles (HFCVs) as well as innovative supporting technologies for variable RE, for example, energy storage systems (ESSs). This requirement of resources and materials occurs over the whole supply chain of the technologies, from the extraction of resources, the manufacture of technology, and the deployment of technology, till the very end of its life cycle. In such context, the consideration of resources in general and critical raw materials (CRMs) in particular and their relations to the risk of supply chain disruption are highly important for achieving the global green energy transition. This editorial paper provides a brief view of the close connectivity between materials/resources and the green transition over the whole supply chain of energy technologies.

The editorial paper includes 11 papers covering the energy transition all over the globe. In these papers, the future national energy transition with a specific energy or climate targets is predicted by applying the energy model [1–3] and relevant energy, materials, and resources required for energy production can be estimated [2,3]. At the global level, [4] study the relations between fossil and renewable resources for energy transition, taking into account the energy security and regional trade. Some authors extend to 'soft' measures for low carbon energy transition such as the energy prosumer business model [5] or the sector coupling of water and energy supply [6,7]. Apart from environmental benefits, the economic, social, and sustainable consequences of RE technologies and energy transition are quantified and assessed [8–10]. A list of CRMs for energy transition and their availability index is presented in [11].

Limpens et al. [1] use the EnergyScope Typical Days model to analyze the Belgian energy system in 2035 for different carbon emission targets. It is a regional, bottom-up and linear model considering multiple sectors and multiple energy carriers with an hourly resolution and a 1-to-5 min computational time. This model optimizes the design and operation strategies of the system including a set of 96 energy technologies, from 24 resources while meeting the end-use demand of electricity (TWh), heat (TWh), mobility (passenger km and tonne km) and non-energy demand (TWh), and minimizing the total annual cost of the system. Besides, the optimization of the system was constrained under a climate target limiting its annual life cycle GHG emissions. It is identified that by 2035, Belgium will lack 275.6 TWh/year of local resources, and 173.3 TWh/year if non-energy demand is not taken into account. To pursue the cost-effective, green energy transition, the demand gap could not be met by individual renewable energy technologies such as offshore wind, geothermal or nuclear power, consequently requiring a mix of renewable solutions. At the same time, the imported renewable fuels or electricity is not a cost-competitive solution (assuming that the price of imported renewable fuels is 50% higher than that of the fossil ones), except for aiming at very low emissions. [1]

Citation: Riva Sanseverino, E.; Luu, L.Q. Critical Raw Materials and Supply Chain Disruption in the Energy Transition. *Energies* **2022**, *15*, 5992. https://doi.org/10.3390/en15165992

Received: 7 August 2022
Accepted: 16 August 2022
Published: 18 August 2022

Rixhon et al. [2] conducted an uncertainty analysis with a whole-energy system model to study the importance of electro-fuels such as hydrogen, methane, and methanol in Belgium's energy transition by 2050. The applied whole-energy system model was EnergyScope Typical Days, which is the same as that of [1]. Only two differences are made, including the timeframe of the study, and the negligence of non-energy end-use demand. In the model, Belgium was modeled as a single node without taking into account intra-national energy transmission. For the uncertainty analysis, the polynomial chaos expansion method was used to highlight the influence of the critical parameters of energy/fuel price, transportation technology costs, technological efficiency, and nuclear power capacity on the total cost of the system.

Under the uncertainties, the annual system cost of 43.6 billion Euros by 2050 could become 17% higher and twice more uncertain in the context of the zero-emission target. Specifically, the price of imported renewable electrofuels is the most critical parameter, contributing to 53.2% of the variation in the total cost of the system. The price of fossil hydrocarbons significantly impacts the variation of the system cost, at 34.8%. The maximum capacity of nuclear power plants has a limited impact of 13.1% on the variation of the total annual cost of the system. Lastly, two transportation-related parameters of the investment cost of cars and of fuel cells have a small impact on the variation of the system cost. The limitation of this uncertainty analysis study lied in the independency of parameters. Though the independency of the parameter is required by the polynomial chaos expansion method, it does not reflect the reality of technology modeling in some cases, for example, the close relation of the technology's investment cost and its efficiency. The authors suggested future studies on the application of electrofuels/biofuels and the characterization such as price, availability, geographical origin, production process, etc of imported electrofuels/biofuels to make the model more refined, realistic and comprehensive. [2]

Delannoy et al. [3] combine GlobalShift and a dynamic function to model the Energy Return On Investment (EROI) of natural gas at a global scale by 2050. GlobalShift composes of data on gas reserves and production for the period of 1950–2050 by gas-producing countries, which is used to quantify energy production. The dynamic function is then applied to analyze the uncertainty of EROI. It is found that 2040 will see the gross energy peak of 249 EJ, while the net energy will reach a peak at 210 EJ in 2037. The average EROI steadily decrease from 141.5 to 16.8 between 1950 and 2050. The energy required to produce gas is 11 EJ, corresponding to 6.7% of gross energy produced in 2020. By 2050, this number will mount up to 53 EJ, or 23.7% of gross energy production. With the exponential increase in the required energy for gas production or the sharp decrease in its EROI, there is a risk to energy security as well as the environment, which suggests the inclusion of EROI in energy transition studies [3].

Berdysheva and Ikonnikova [4] propose a modified index for energy security, and apply it to the global energy trade to understand the growth in the unconventional resources in the United State of America, RE in Europe (EU), Chinese natural gas consumption, and changes in other countries' energy flows, as well as their relations to the energy transition, the economic situation and the trade network. The authors use a six-step approach of (1) update data on energy production, consumption, and trade 2000–2008, (2) compare data of the International Energy Agency and United Nations' commodity trade to see the energy flows, and (3) compile data for monetary flows to see the economic link, (4) characterize individual -economy energy systems' evolution, in relation to trade, (5) apply complex network method to see the evolution of trade and test the small world property to see the change in the cluster and network of energy over time, and (6) use modified energy security index to see the change in demand, supply, and trade. The results show that the green energy transition toward higher investment in RE does not improve energy security in most countries (even make it worse). The reduction in coal consumption changes the fuel diversification balance and weakens energy security. The increased reliance on natural gas causes a negative impact on energy security; but expanding the liquified natural gas trade reduces the negative impact. The growth in global energy demand induces major energy exporters to produce more, exposing them to supply risk.

The business model of tenant electricity, which provides tenants of a building with on-site solar power, offers the potential to achieve energy transition and GHG reduction targets. In Germany, Moser et al. [5] study barriers to and drivers of diffusion of the tenant electricity, using qualitative data analysis and semi-structure expert interviews. The identified main barriers are the legal framework which causes high transaction costs, and the reluctancy of residents to become prosumers of electricity. Meanwhile, the drivers of this business model include increasing electricity demand, technical development such as blockchain and smart meters, and EU RE directives [5].

Torabi et al. [6] study the penetration level of electric vehicles (EVs), and sector coupling (of water supply and energy management) in an island of Portugal to highlight the contribution of optimized management of RE resources on its energy transition. The island's energy system is transitioning towards the dominance of solar and wind energies. With the high share of RE, curtailment is inevitable. To support this transition and minimize the curtailment, three solutions have been identified, including the deployment of ESSs, EVs, and demand side management of water desalination plants. These solutions are evaluated by optimizing the system while maintaining the power supply being equal to the demand plus curtailed power. It was identified that the share of RE may reach 100% and the curtailment events could be reduced by the large-scale deployment of EVs and demand management of desalination plants and charging management of ESSs and EVs. At the same time, the greenhouse gas emissions of the mixed grid reduce accordingly [6].

Zohrabian and Sanders [7] estimate the energy and GHG emission trade-offs of projected water supply in Los Angeles by 2050. The electricity demand for surface water supply and recycled water system between 2010 and 2050 is calculated by applying an energy intensity for annual water volume from different sources. The factors impacting electric demand for water supply are then decomposed to highlight their importance. The corresponding GHG emissions are quantified with the current emission intensity of the current and future grid mix. The results show that treating stormwater and recycling water bring benefits for coping with water shortage; however, these measures might not considerably benefit in terms of electricity demand. Water conservation brings benefits of energy savings which are higher in the case of using locally supplied water than imported water. At the same time, increasing the local water sources in replacement of imported water will cause the geospatial change in energy demand from outside the city (for recycling water) to inside the city (for pumping local water). As a result, the local electricity system and its corresponding GHG emissions will be impacted. The decomposition analysis indicates that the change in the local water supply structure has a higher impact on the electricity demand than population growth and water conservation [7].

Bethoux [8] studied the barriers to expanding the deployment of Hydrogen Fuel Cell Vehicles, HFCVs, on the mass road transportation vehicle market, considering the environmental and economic aspects over the whole supply chain of production, storage, and distribution of hydrogen. It is identified that there is a market for using hydrogen for both light and heavy road transportation. Green hydrogen may be one of the potential uses of renewable energies and natural hydrogen might become an economic reality pushing the HFCVs to be a competitive and environmentally friendly alternative to battery electric vehicles. In the meantime, some barriers that need to be overcome, so as to reduce the vehicle and fuel technologies' cost, increase vehicle durability, the lower environmental footprint of the vehicle, especially in the manufacturing and disposal stages, improve hydrogen production technologies, enhance the safety of the hydrogen infrastructure as well as the vehicle [8].

Pietrzak et al. [9] conducted a critical situation assessment of RE sources in Poland, taking into account three aspects of physical energy sources, energy policy, and social awareness. Through a semi-structured expert-assessment survey, the study points out that Poland has large RE resources, and there is a potential for further exploiting this resource in the near future. Specifically, the potentials for solar, wind, and solid biomass development are assessed to be the highest among different RE technologies. Some factors preventing the deployment of RE have been identified, including the conventional energy lobby, complex RE regulation, and high investment costs. In order to achieve the energy transition, five activities such as change of the national law, public education on RE, financial incentive and tax exemption for RE investment, development of prosumer energy, and dialogue with the coal lobby are suggested by the experts [9].

Hale and Long [10] evaluate sustainability outcomes of energy transition using univariate time series prediction model. The authors use exponential smoothing and AutoRegressive Integrated Moving Average (ARIMA) model to predict the annual electricity generation supply by 2029. The predicted electricity generation with the lowest uncertainty, obtained with the ARIMA model is assessed for four sustainability indices of carbon, water, land, and cost footprints. The change in electricity generation structure (reduction in coal and increase in solar and wind) and the increase in electricity generation during 2020–2029 will cause an increase in land and cost footprints, but a decrease in carbon and water footprints. In case the increase in coal-based electricity is substituted by solar only, the land footprint increases by the smallest rate, and but the cost is the largest among different substitution strategies. Meanwhile, the substitution with wind is the best strategy in terms of water and cost footprint, but the worst one in terms of land footprint [10].

Nate et al. [11] provide an availability scoring of 17 critical materials concerning 10 energy technologies. The availability of these critical materials is ranked by their current, absolute amount used for energy technologies, their projected, percentage annual demand by 2050 compared to the current value, the number of technologies requiring these critical materials, their accumulative emissions of CO2, their reserves availability, the number of countries producing more than 1% of global production and the countries with highest annual material productivity. Two supply-demand scenarios have been developed using independent parameter probability and supply-demand balanced fuzzy estimation. It is identified that cobalt, graphite, and lithium, which are used for ESSs, have the lowest material availability ranking index. These materials are followed by iron, nickel, and chromium. With the changes in the supply-demand balance, cobalt, lithium, rare earth elements, iron and vanadium are the most unpredictable materials [11].

Author Contributions: Writing—original draft preparation, E.R.S. and L.Q.L.; writing—review and editing, E.R.S. and L.Q.L. All authors have read and agreed to the published version of the manuscript.

Funding: This research received no external funding.

Conflicts of Interest: The authors declare no conflict of interest.

References

1. Limpens, G.; Jeanmart, H.; Maréchal, F. Belgian Energy Transition: What Are the Options? *Energies* **2020**, *13*, 261. [CrossRef]
2. Rixhon, X.; Limpens, G.; Coppitters, D.; Jeanmart, H.; Contino, F. The Role of Electrofuels under Uncertainties for the Belgian Energy Transition. *Energies* **2021**, *14*, 4027. [CrossRef]
3. Delannoy, L.; Longaretti, P.-Y.; Murphy, D.J.; Prados, E. Assessing Global Long-Term EROI of Gas: A Net-Energy Perspective on the Energy Transition. *Energies* **2021**, *14*, 5112. [CrossRef]
4. Berdysheva, S.; Ikonnikova, S. The Energy Transition and Shifts in Fossil Fuel Use: The Study of International Energy Trade and Energy Security Dynamics. *Energies* **2021**, *14*, 5396. [CrossRef]
5. Moser, R.; Xia-Bauer, C.; Thema, J.; Vondung, F. Solar Prosumers in the German Energy Transition: A Multi-Level Perspective Analysis of the German 'Mieterstrom' Model. *Energies* **2021**, *14*, 1188. [CrossRef]
6. Torabi, R.; Gomes, Á.; Morgado-Dias, F. Energy Transition on Islands with the Presence of Electric Vehicles: A Case Study for Porto Santo. *Energies* **2021**, *14*, 3439. [CrossRef]
7. Zohrabian, A.; Sanders, K.T. The Energy Trade-Offs of Transitioning to a Locally Sourced Water Supply Portfolio in the City of Los Angeles. *Energies* **2020**, *13*, 5589. [CrossRef]
8. Bethoux, O. Hydrogen Fuel Cell Road Vehicles and Their Infrastructure: An Option towards an Environmentally Friendly Energy Transition. *Energies* **2020**, *13*, 6132. [CrossRef]
9. Pietrzak, M.; Igliński, B.; Kujawski, W.; Iwański, P. Energy Transition in Poland—Assessment of the Renewable Energy Sector. *Energies* **2021**, *14*, 2046. [CrossRef]

10. Hale, J.; Long, S. A Time Series Sustainability Assessment of a Partial Energy Portfolio Transition. *Energies* **2020**, *14*, 141. [CrossRef]
11. Nate, S.; Bilan, Y.; Kurylo, M.; Lyashenko, O.; Napieralski, P.; Kharlamova, G. Mineral Policy within the Framework of Limited Critical Resources and a Green Energy Transition. *Energies* **2021**, *14*, 2688. [CrossRef]

energies

MDPI

Article

Examining the Impact of Energy Price Volatility on Commodity Prices from Energy Supply Chain Perspectives

Hokey Min

Maurer Center 312, Allen and Carol Schmidthorst College of Business, Bowling Green State University, Bowling Green, OH 43403, USA; hmin@bgsu.edu; Tel.: +1-419-372-3442

Abstract: Oil has historically been the most significant primary energy source for our daily lives and business activities. However, recent skyrocketing oil prices have been one of the greatest concerns among policymakers, business executives, and the general public due to their impacts on daily necessities, including food, clothing, and automobile transportation. As a result, fast-rising inflation on the global scale is attributed to mounting oil prices. Even though many countries have made a conscious effort to tame oil prices and the subsequent inflation, their efforts are often in vain due to some uncontrollable situations. These situations include the ongoing war between Ukraine and Russia, where Russia began weaponizing its oil resources and limiting oil supplies to its neighboring European countries. Faced with the current energy crisis, a growing number of policymakers and business executives have attempted to develop energy-induced risk mitigation strategies. With this in mind, the primary purpose of this paper is to investigate what may have caused oil price hikes and to determine how significantly oil prices influence commodity prices. This paper then proposes ways to mitigate energy-induced supply chain risks by analyzing four decades of secondary data obtained from multiple sources.

Keywords: energy price volatility; energy supply chain; commodity pricing; supply chain mapping; supply chain resilience; secondary data analysis; trend analysis

Citation: Min, H. Examining the Impact of Energy Price Volatility on Commodity Prices from Energy Supply Chain Perspectives. *Energies* **2022**, *15*, 7957. https://doi.org/10.3390/en15217957

Academic Editor: Audrius Banaitis

Received: 3 October 2022
Accepted: 24 October 2022
Published: 26 October 2022

1. Introduction

For the last two years, global oil prices have surged, with benchmark Brent crude jumping from an average of USD 41.96 per barrel at the peak of the COVID-19 pandemic in 2020 to USD 107.64 in July 2022 [1]. The International Energy Agency expects the current energy demand to increase by 37% in the next 25 years [2]. Similarly, the International Energy Outlook 2019 [3] predicts significant growth in worldwide energy demand over the 28 years from 2012 to 2040. Total global energy consumption is expected to grow from 549 quadrillion British thermal units (Btu) in 2012 to 629 quadrillion Btu in 2020 and eventually to 815 quadrillion Btu in 2040—a 48% increase from 2012 to 2040 [4]. In particular, U.S. gasoline and diesel inventories are running low, refining capacity is constrained, and oil demand remains strong due in part to the pent-up demand [5]. A massive increase in energy prices puts a heavy burden on every household, with soaring costs for electricity and mobility. Subsequently, rising energy (especially oil) prices have created unprecedented economic crises worldwide through superinflation. Energy price hikes are primarily caused by rapidly growing worldwide demand for oil in the wake of extreme weather conditions, a shortage of oil supply to the European Union (E.U.) from Russia, and a lack of investment in energy grid stability due to austerity government policies following the COVID-19-induced economic doldrums. In particular, the E.U. nations' dependency on the imports of fossil fuels from Russia, which is using oil as its geopolitical weapon, poses serious additional challenges in defusing the current energy crisis. Faced with these challenges, policymakers and business executives need to act immediately to change their energy conservation policies and strategies. These polices include oil rationing, energy supply diversity, and energy supply chain security. Another change in business strategies includes the overhaul of private-sector energy usage practices, such as industrial energy curtailment. Cases in point, the state of California has already issued electricity outage warnings and imposed restrictions on air conditioning usage. Texas, Illinois, and Missouri will likely develop energy policies (e.g., rolling blackouts) similar to those in California amid sweltering summer heat in 2022.

Petroleum (crude oil) is a fossil fuel that is a non-renewable source of energy. Although crude oil (oil hereafter) is a primary energy source for generating electricity, gasoline, and heating, oil is a significant culprit of air pollution and the subsequent greenhouse effect since burning oil emits carbon dioxide gas and contributes to climate change through global warming. To make matters worse, a vast majority of countries worldwide have imported oil from the Organization of the Petroleum Exporting Countries (OPEC) and Russia at increasingly high prices as worldwide oil reserves shrink. In particular, electricity generated from oil is costly compared to other fossil fuels, such as coal and gas.

2. Relevant Literature Review

Despite the inherent complexity and volatility of energy prices, an accurate forecast of energy prices and an understanding of energy price behaviors would help energy producers and consumers determine their energy production capacity, energy inventory level, and size of investments in energy generation and distribution. Such energy price forecast can help the energy user or energy developer select the most cost-efficient energy sources. Given the significance of energy pricing to economic activities, some scholars have attempted to capture energy pricing patterns and assess their impacts on economic activities. To elaborate, Asafu-Adjaye [6] was one of the first to estimate the causal relationships between energy consumption and income in India, Indonesia, the Philippines, and Thailand, using co-integration and error-correction modeling techniques. His study result indicates a causal relationship between energy prices and income. Finn [7] theorized that energy price shocks equivalent to adverse technology shocks could induce significant contractions in economic activity. Brown and Yücel [8] found that oil price shocks could affect aggregate economic activities. Thus, they argued that both monetary and energy policies should be developed based on energy price fluctuations.

Similarly, Papapetrou [9] observed that oil price changes affected actual economic activity and employment. Oil prices were found to be important in explaining stock price movements based on the empirical evidence obtained from a multivariate vector autoregression (VAR) approach. His study, however, was confined to Greece. Huang et al. [10] applied the multivariate threshold model to investigate the impacts of an oil price change and its volatility on economic activities (changes in industrial production and actual stock returns) and found that an oil price change or its volatility had a limited impact on the economies if the change was below threshold levels. If the change was above threshold levels, an oil price change or volatility affected economic activities more significantly than the real interest rate. Their data, however, were limited to monthly data from the US, Canada, and Japan from 1970 to 2002.

However, Olomola [11] found that oil price shocks did not affect economic output or inflation in Nigeria, while oil price shocks significantly influenced the actual exchange rates in Nigeria. Benkraiem et al. [12] investigated the relationship between S&P 500 prices as a U.S. economic barometer and a set of energy prices, including WTI crude oil prices. They observed that crude oil price shocks influenced short- and long-term U.S. stock market dynamics. Balashova and Serletis [13] discovered that oil price shocks had a positive and statistically significant impact on almost all types of Russian economic activity, including the economic output of manufacturing, mining, construction, transport, retail, and wholesale trade.

Similar to the above line of research examining the causality between oil price shocks and economic activity, Carfora et al. [14] recently examined the causal relationships among energy prices, income, and energy consumption in selected Asian countries (India, Indonesia, Thailand, and the Philippines). Although those relationships varied from one country to another, they found that, in the cases of India and Indonesia, a bidirectional relationship existed between energy prices and income. Dagoumas et al. [15] re-investigated the long-run relationship between energy prices and economic growth within the periphery of the European Union (E.U.) using the Variance Decomposition Analysis. Given that energy prices were strong drivers of inflation in the E.U., they examined how the energy prices (especially crude oil prices) affected economic growth within the E.U. periphery. They found that energy prices negatively affected Gross Domestic Product (GDP) growth rates in the E.U.

As this review of the prior literature on energy pricing reveals, most of the existing studies on energy pricing focused on the macroeconomic implications of energy pricing. Refocusing on the microeconomic implications of energy pricing, some pioneering works started to investigate how energy prices affected commodity prices sensitive to economic fluctuations and, thus, better reflect economic conditions. To elaborate, after recognizing an increase in the interconnections of agriculture and energy markets through the rise in new biofuel agribusinesses and oil–ethanol–corn linkages, Saghaian [16] reported a strong correlation between oil and commodity prices based on empirical results obtained from the contemporary time-series analysis and Granger causality. Following up, Koirala et al. [17] examined whether linear relationships exist between future energy prices and future prices of agricultural commodities, such as corn and soybeans. Their study results revealed that future agricultural commodity and energy prices were highly correlated; thus, an increase in energy prices increased the prices of corn and soybeans. Concerned about a surge in agricultural commodity prices in South Africa from 2004 to 2008, Fowowe [18] analyzed the relationship between oil prices and commodity prices and found no evidence that agricultural commodity prices in South Africa responded to oil prices. This finding contrasted with the prior findings of the earlier studies. López Cabrera and Schulz [19] investigated price and volatility risk originating in linkages between energy (especially biofuel) and agricultural commodity prices in Germany using an asymmetric dynamic conditional correlation GARCH model, as well as a multivariate multiplicative volatility model. Similar to Fowewe [18]'s study finding, they found that the long-run correlation between energy (biodiesel) prices and agricultural commodities (rapeseed) was relatively low and not significant. They also noted that biodiesel prices did not influence rapeseed and crude oil prices in the short run. In contrast, Wei et al. [20] confirmed a bidirectional positive causality between oil and agricultural commodity prices. These earlier studies focused on examining any causality between oil and agricultural commodity prices under the premise that oil price hikes would lead to a greater use of alternative energy (especially biofuel) extracted from agricultural commodities (e.g., corn and rapeseed) and to an increase in agricultural commodity and food prices.

Considering the shortcomings mentioned above and the paucity of earlier studies on energy pricing implications, this paper analyzes more than four decades of secondary data regarding global oil prices; gasoline prices; and beef, pork, cotton, gold, silver, iron ore, and coffee prices. Furthermore, this paper examines any functional connection between global oil price volatility and commodity prices. This paper also discusses various managerial or policy implications of energy price volatility from an energy supply chain perspective.

3. Sources of Energy Crisis

Given energy's direct impact on our costs of living, such as electricity and heating bills, many households and enterprises pay close attention to energy pricing and wonder why energy pricing is so volatile and thus difficult to predict. The volatility of energy prices is attributed to a multitude of complicated factors. In particular, since oil has been a primary energy source, I focus on identifying the main factors influencing oil price volatility. These factors include the following:

(1) Demand for Crude Oil: Volpe [21] recently reported that, based on the data available from the American Petroleum Institute (API), the prices of gasoline are often determined by the cost of global crude oil (61%), refining costs (14%), distribution and marketing costs (11%), and federal and state taxes (14%). Therefore, it is apparent that the price of a barrel of crude oil in open markets dictates the price of fuel that people consume every day. Due to the variety and different blends of crude oil, its

price depends on one of the four popular benchmarks: Brent Crude, West Texas Intermediate (WTI), Dubai Crude, and OPEC baskets. Brent Crude is the most widely used one and is typically sold on the spot market at London's International Petroleum Institute. At the same time, WTI is the U.S. benchmark for light sweet oil traded on the New York Mercantile Exchange (NYMEX) for gasoline. The Dubai Crude (called Fateh) represents a medium sour crude oil extracted from Dubai. Dubai Crude is used for pricing Persian Gulf crude oil exports to Asia [22]. The OPEC basket price is a weighted average of the prices of 13 regional oils from Algeria's Saharan Blend, Angola's Girassol, Ecuador's Oriente, Indonesia's Minas, Iran's Heavy, Iraq's Basra Light, Kuwait's Export, Libya's Es Sider, Nigeria's Bonny Light, Qatar's Marine Saudi Arabia's Light, the United Arab Emirates' Murban, and Venezuela's Merey [23]. To complicate oil pricing, the benchmark mentioned above can be determined through either the spot market or future prices. Two contrasting market situations can set future prices: (1) Backwardation, where market prices are expected to be lower in the future months than the present day, and (2) Contango, where market prices are expected to be higher in the future months than the present day.

(2) Government Policy, Regulations, and Laws: U.S. gasoline is subject to federal and state taxes. As of 2022, U.S. federal taxes consisted of excise taxes of USD 0.183 per gallon on gasoline, USD 0.243 per gallon on diesel fuel, and a leaking underground storage tank fee of USD 0.01 per gallon on both fuels [21]. This fact illustrates that fuel price is affected by the government's tax policy. In addition, since oil drilling and production can be regulated by state laws in the U.S., oil supplies and subsequent changes in oil market pricing controlled by the state government can affect oil pricing. The U.S. federal government regulates offshore oil exploration for the Outer Continental Shelf (a radius of 200 nautical miles offshore) and thus influences oil production and pricing. Furthermore, stricter government regulations (e.g., Environmental Protection Agency regulations) intended for environmental protection can hurt oil pricing. Not to mention the U.S. policies, the OPEC policies regarding its oil production tend to have a profound impact on global oil prices since OPEC accounts for 40% of the world crude oil production, and its oil exports represent about 60% of the total petroleum traded globally [24]. Another example is Venezuela and Nigeria's nationalization of oil fields, which led to global oil shortages and price increases soon after those countries' abrupt policy shifts.

(3) Political Instability, Unrest, Geopolitical Tension, and War: Historically, civil uprisings, changes in political power, border conflicts, and regional wars involving oil-producing countries disrupted oil supplies and created a ripple effect on oil prices. For instance, the Gulf War in the early 1990s, triggered by Iraq, caused a 9-month oil price hike and nearly doubled oil prices [25]. Similarly, the ongoing war between Ukraine and Russia has led to a series of import bans for Russian crude oil, liquefied natural gas, and coal by the U.S. and European Union (E.U.), subsequently increasing global oil prices in the year 2022. Indeed, the price of crude oil in the global market skyrocketed from approximately USD 76 per barrel at the start of January 2022 to over USD 110 per barrel in March 2022 due to Ukraine and Russia's border conflicts [26].

(4) Natural Calamities and Disasters: Natural disasters, such as hurricanes, tornadoes, flooding, earthquakes, and tsunamis, can wreak havoc on energy infrastructure, including oil refineries and power plants. For example, when Hurricane Katrina hit the U.S. Gulf Coast region, which accounted for 35% of oil production in 2005, U.S. oil prices soared by around 20% [27]. However, when a 9.0-magnitude earthquake rocked Japan in 2011 and then destroyed six oil refineries that accounted for 31% of Japan's oil output, many expected a temporary oil price drop since refinery closures would result in reduced crude oil imports [28].

(5) Trader's Speculative Investment in Oil: Generally, when crude oil supply is tight or it is considered valuable (premium), its price goes up, whereas if its demand is low or it is considered less valuable, its price decreases due to its discount. In the NYSE,

oil traders determine the volume of speculative crude oil purchases and thus affect the overall demand for crude oil and the subsequent future oil prices. To meet the U.S. Renewable Fuel Standard (RFS) program targets, the EPA also requires U.S. oil companies to have one Renewable Identification Number (RIN) for each gallon of ethanol blended into fuel [29]. To comply with this requirement, some oil producers that are RIN-short need to increase the purchase of RIN (e.g., biodiesel fuels). Thus, their oil trade volume can affect oil prices. Furthermore, some industries (e.g., airline and trucking sectors) participating in cooperative hedging programs against fuel price hikes can increase their speculative investments in crude oil and the subsequent oil price, especially when many companies speculate towards a continued upward pressure on oil prices.

(6) Grid Network, Power Generation, and Distribution: Tayeb [30] recently reported that U.S. power grid failures in most of Western and Central U.S. increased vulnerability to the energy supply chain and increased the risk of electricity shortfalls and disruptions. With rising demand for additional power generation, the U.S. government has been under growing pressure to expand the power grid. However, adding high-voltage transmission lines and switches to the grid usually takes much time. In contrast, replacement parts for turbines and other equipment needed for the power grid may not be readily available. In addition, power plant commissioning delays can aggravate the grid network problem. An obsolete and insufficient grid network can adversely affect oil prices.

(7) Alternative Energy Availability and Affordability: Recognizing the mounting cost of using fossil fuels and their contribution to global warming, a growing number of energy producers, including power plants, are exploring various sources of alternative energy. These include solar, wind, geothermal, biomass, hydrogen, tide/wave, natural gas, municipal waste, coal, and nuclear. All of these alternative energy sources, apart from coal, are clean or renewable energy sources. In particular, since renewable energy can derive power from natural sources, it can replenish itself without running out. Due to such benefits, the use of alternative energy has grown exponentially in recent years, accounting for 23.2% of all energy sources for power generation in 2020 [31]. The International Energy Agency (IEA) predicted that alternative energy sources would account for nearly half of the worldwide increase in power supply up to 2040 [32]. The increased use of alternative energy will decrease oil demand and lower oil prices.

(8) Energy Waste: According to the Energy Information Administration (EIA), two-thirds (66%) of the primary energy used to create electricity is wasted by the time the electricity arrives at the customer's meter. Generally, more than half (59%) of energy is lost in the power generation process due in part to waste heat released in the air and inefficient transformers and equipment, including pumps, fans, and industrial boilers [33]. If energy waste can be reduced, energy consumption will drop, thus decreasing oil prices.

As discussed above, there exists a host of factors influencing oil prices. Though not specified, other factors, such as inflation and currency fluctuations, can contribute to oil price volatility. Due to complicated oil price volatility, it is challenging to forecast oil prices and assess their impacts on our standard of living and daily business practices. Recognizing such a challenge, the primary purpose of this paper is to examine any functional relationship or link between oil price volatility and commodity pricing that shapes our daily lives and everyday business practices. This paper proposes a series of hypotheses and tests them using statistical data analyses, including a regression analysis and a trend analysis, predicated on more than four decades of various pricing data collected from secondary sources.

4. Propositions, Analyses, and Results

Due to the volatility and complexity of crude oil pricing, it is a daunting task for us to accurately predict future oil prices and to assess their potential impacts on commodity prices. To understand oil pricing dynamics and their ramifications for commodity markets, I experimented with multiple business analytic tools (e.g., a series of statistical data and forecasting analyses) with secondary data obtained from multiple public sources. These sources include IEA's Energy Statistics Data Browser, Nasdaq Data Link, Refinitiv Eikon-Commodities Data Catalogue, Internal Monetary Fund's Commodity Data Portal, World Bank Commodity Prices Database, and Wall Street Commodity Data. The following subsections provide details of those experimental results and their managerial implications.

4.1. Experimental Data

I collected monthly time-series data about the prices of popular energy sources, comprising crude oil, gasoline, diesel, and Austrian coal from the secondary data sources that I referred to earlier. I also compiled matching data about the prices of selected commodities: (1) metals, such as aluminum, gold, silver, and iron ore; (2) agricultural commodities, such as corn, cotton, coffee, and wheat; and (3) meats, such as beef and pork. The data set contained 449 monthly pricing records for 37 years, from March 1985 to July 2022. I compiled the collected data into formats of both Excel® and SPSS files for statistical data analyses.

4.2. Propositions

When the price of crude oil rises, people have to decide how often they should travel, how often they should go grocery shopping, or how much they should spend without going over their budgets since oil price hikes tend to impact people's mobility, heating/electricity bills, and subsequent daily spending. In particular, Americans' daily lives are heavily dependent on oil, as they are the biggest oil consumers in the world. The United States uses 20.54 million barrels of oil daily, accounting for approximately 20% of the 100.23 million barrels produced daily worldwide [34]. Although people have long felt the impact of oil prices on their livelihood, few scientific studies have examined the correlation between oil prices and the cost of living. With that in mind, I developed a series of propositions that test the validity of relationships between oil price volatility and the cost of living reflected in the prices of commodities that are essential for sustaining our daily lives. These commodities include wheat, corn, coffee, beef, and pork, which comprise ordinary people's daily food menu. In addition, I included other commodities, such as cotton, an essential material for clothing; aluminum and iron, which represent essential ingredients for many products (e.g., automobiles); and gold and silver, which represent popular investment targets as currency replacements. Furthermore, I added coal since it can be substituted for oil as an alternative energy source.

Proposition 1. *There is a positive relationship between oil and corn prices.*

Based on the premise that crude oil prices can increase conventional fossil fuel (e.g., gasoline) prices, I propose that oil price hikes will increase the demand for alternative energy, such as biofuel (e.g., ethanol), which can be created from corn, consequently increasing corn prices. As of 2009, corn use for ethanol accounted for approximately one-third of the total demand for U.S. corn [35]. However, corn processing for ethanol will continue to proliferate in the next few years with government mandates calling for increased ethanol use in the wake of sky-high oil prices. Such growth is likely to further increase corn prices. When this proposition is tested using a correlation analysis, corn price has a significant positive correlation with oil price with a Pearson correlation coefficient of 0.811 ($p = 0.000$). Figure 1 graphically shows the positive correlation between oil and corn prices for the last four decades. As expected, I also found that both gasoline and diesel prices have strong positive correlations with corn prices. Specifically, gasoline price significantly correlates with corn price with a Pearson correlation coefficient of 0.832. Likewise, diesel price significantly correlates with corn price with a Pearson correlation coefficient of 0.715. In addition, I performed a regression analysis to conduct an inference test with the corn price as the dependent variable and the crude oil price as the independent variable.

This test result also confirms that crude oil is a significantly good predictor of corn price (with a standardized β coefficient of 0.811 and an adjusted R-square value of 0.657).

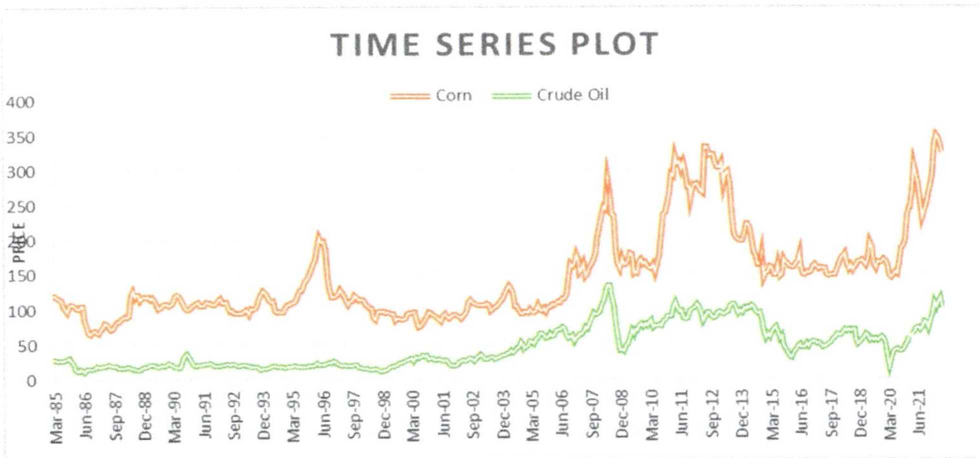

Figure 1. Time-Series Plot of Crude Oil and Corn Price Patterns.

Proposition 2. *There is a positive relationship between oil and coal prices.*

Using the logic similar to proposition 1, I hypothesize that oil price hikes will increase the demand for another alternative energy source, such as coal, thus increasing coal price. The test result of this proposition reveals that coal price is positively related to oil price, as illustrated in Figure 2. The correlation analysis shows that coal price has a strong positive correlation with oil price, with a Pearson correlation coefficient of 0.831 ($p = 0.000$). The inference test based on a bivariate regression analysis with the coal price as the dependent variable and the crude oil price as the independent variable shows that crude oil is a good predictor of coal price (with a standardized β coefficient of 0.831 and an adjusted R-square value of 0.690).

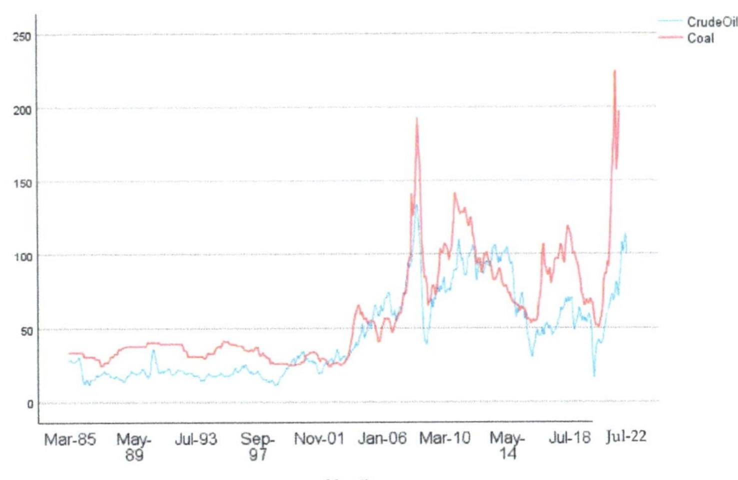

Figure 2. Time-Series Plot of Crude Oil and Coal Price Patterns.

Proposition 3. *There is a positive relationship between oil and iron ore prices.*

As the world's second most traded bulk commodity, iron ore is one of the essential minerals needed for producing industrial goods, such as machinery, tools, vehicles, aircraft, ships, building structures, and bridges. Iron ore production is dominated by Brazil and Australia, which accounts for 80% of iron ore supplies [36]. Since the world's largest iron ore consumer is China, importing iron ore from remotely located Brazil and Australia to China will likely involve bulk shipping affected by fuel cost. Since fuel cost is dictated by oil price, the rising oil price will lead to higher iron ore prices. Under such a premise, I posit a hypothesis that iron ore price is directly related to oil price. This hypothesis is validated in that iron ore price has a strong positive correlation with oil price, with a Pearson correlation coefficient of 0.787 (p = 0.000). Figure 3 graphically displays the matching pricing patterns of crude oil and iron ore. The regression analysis result with the iron ore price as the dependent variable and the crude oil price as the independent variable shows that crude oil is a good predictor of iron ore price (with a standardized β coefficient of 0.787 and an adjusted R-square value of 0.619).

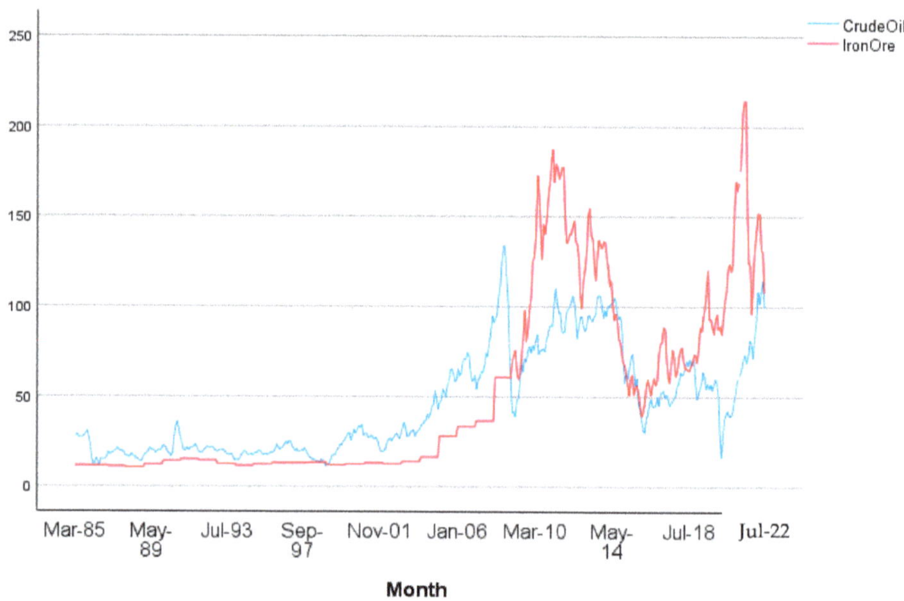

Figure 3. Time-Series Plot of Crude Oil and Iron Ore Price Patterns.

Proposition 4. *There is a positive relationship between oil and aluminum prices.*

The aluminum price reached USD 2830 per metric ton (M.T.) in May 2022, the highest price since the two-year high in 2018 [37]. A constant rise in aluminum prices is a cause for concern due to its impact on the cost of production of industrial goods. Kumar [38] recently observed that the hike in aluminum price was attributed to a substantial increase in energy costs, growing demand, a decline in China's production capacity, a reduction in global inventories, and the impact of COVID-19. He also noted that the cost of electricity powered by oil comprised approximately one-third (38%) of aluminum production cost. Therefore, I hypothesize that oil price affects aluminum price. As expected, aluminum price positively correlates with oil price, with a Pearson correlation coefficient of 0.657. Considering that the unit of measurement for aluminum price is substantially higher than that for crude oil price, I transformed the price scales of aluminum and crude oil into logarithmic price scales to represent equivalent price fluctuations on the same vertical scale. This transformation intends to reduce the variance in price scales by making the data conform to the lognormal law of error for inferential purposes [39]. Figure 4 shows the similar time-series patterns of the logarithmic price scales of aluminum and crude oil. The regression analysis result with aluminum price as the dependent variable and crude oil price as the independent variable confirms that crude oil is a good predictor of aluminum price (with a standardized β coefficient of 0.657 and an adjusted R-square value of 0.430).

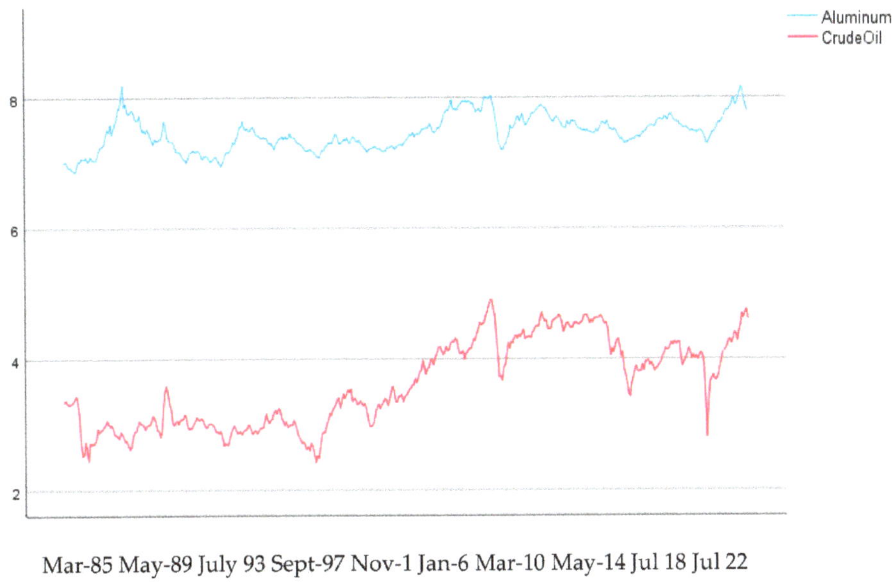

Figure 4. Time-Series (Logarithmic) Plot of Crude Oil and Aluminum Price Patterns.

Proposition 5. *There is a positive relationship between oil and gold prices.*

Over the last half-century, oil prices seemingly fluctuated in parallel to gold prices. Indeed, gold prices rose along with oil prices in the 1970s and 2000s, while both prices dropped simultaneously in the 1980s and 1990s. Based on this observation, some suggest that oil price may drive gold's price, whereas others discount the relationship since the inflationary trend may raise both prices simultaneously [40,41]. To complicate the gold and oil relationship myth, gold is often regarded as a monetary asset (or currency) rather than a typical commodity. Although gold price may be influenced by many dynamic factors, such as inflation, interest rate, and currency (e.g., dollar) valuation, I still found a positive correlation between gold and oil price fluctuations, with a Pearson correlation coefficient of 0.748 (p = 0.000). The regression analysis result with gold price as the dependent variable and crude oil price as the independent variable reaffirms that crude oil is a good predictor of gold price (with a standardized β coefficient of 0.748 and an adjusted R-square value of 0.559).

Even though oil price may be slightly more volatile than gold price as shown in Figure 5, Figure 5 indicates a similar movement pattern for both prices (especially in the 2000s and 2010s). In addition, when I made a similar premise for a potential relationship between silver and oil prices, I still found a significant relationship between oil and silver prices, with a Pearson correlation coefficient of 0.640. That is, gold, silver, and oil prices tended to move together most of the time during the last four decades, and, thus, oil prices can be a predictor of both gold and silver prices.

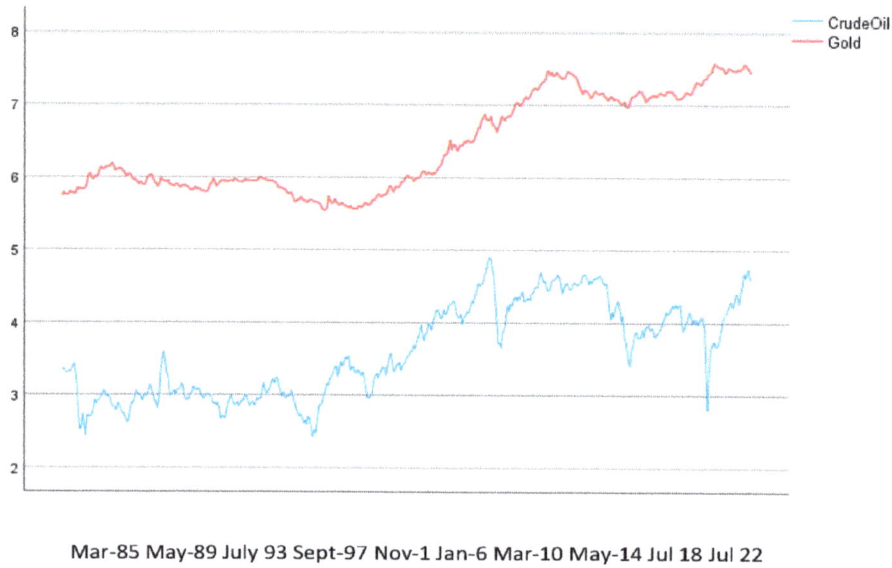

Figure 5. Time-Series (Logarithmic) Plot of Crude Oil and Gold Price Patterns.

Proposition 6. *There is a positive relationship between oil and wheat prices.*

Baffes [42] estimated that grain prices increased 0.18 percent for every 1 percent increase in the price of oil. Cartwright and Riabko [43] discovered that future wheat and oil prices were correlated but not causally related. However, Reboredo [44] found weak oil–food dependence and no extreme market dependence between oil and food prices, including wheat prices, based on a weekly data analysis from January 1998 to April 2011. Given that fuel created by oil is required to run agricultural equipment and process, store, and transport agricultural commodities (such as wheat), crude oil is a critical input to agricultural production. Therefore, I surmise that wheat price may increase with oil price. This premise turns out to be true since wheat price has a strong positive correlation with oil price, with a Pearson correlation coefficient of 0.814 ($p = 0.000$). Although wheat price looks more volatile than oil price, Figure 6 shows a similar price pattern for both wheat and oil prices. The regression analysis result with wheat price as the dependent variable and crude oil price as the independent variable reaffirms that crude oil is a good predictor of wheat price (with a standardized β coefficient of 0.814 and an adjusted R-square value of 0.662).

Proposition 7. *There is a positive relationship between oil and coffee prices.*

Coffee is a tropical commodity that the Commodity-Dependent Developing Countries (CDCs) located in sub-Saharan Africa, South Asia (e.g., India), and Latin America (e.g., Brazil and Columbia) mainly produce. These CDCs are vulnerable to oil price hikes and subsequent supply chain disruptions due to an inadequate transportation infrastructure. Thus, conventional wisdom indicates that the volatility of coffee prices would parallel that of oil prices. Maurice and Davis [45] found a long-run causality between oil and coffee prices. However, Vijayakumar [46] did not find any concrete evidence indicating a correlation between oil and Indian coffee prices. Congruent with the finding of Vijayakumar [46], I found no significant correlation between oil and coffee (especially Robusta Coffee) prices for the last four decades, as shown in Figure 7. In particular, except for in the early 2000s, coffee and oil prices rarely moved in the same direction.

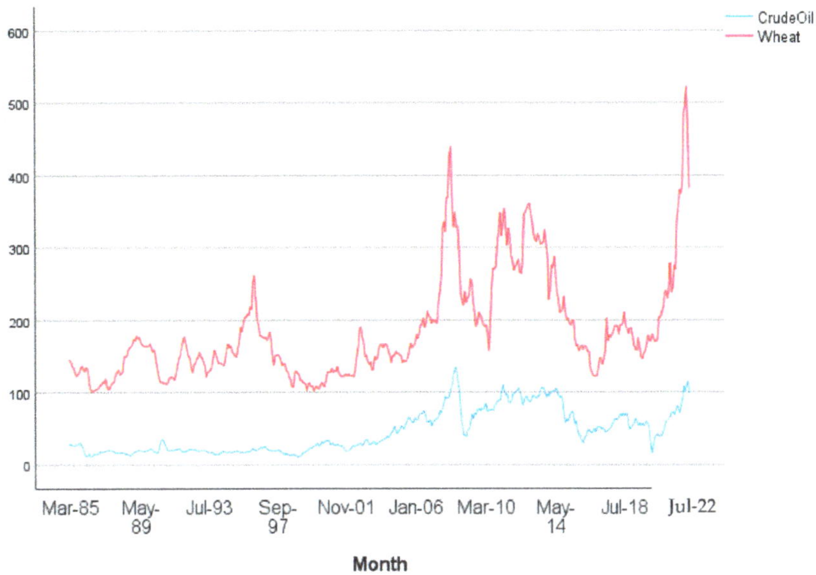

Figure 6. Time-Series Plot of Crude Oil and Wheat Price Patterns.

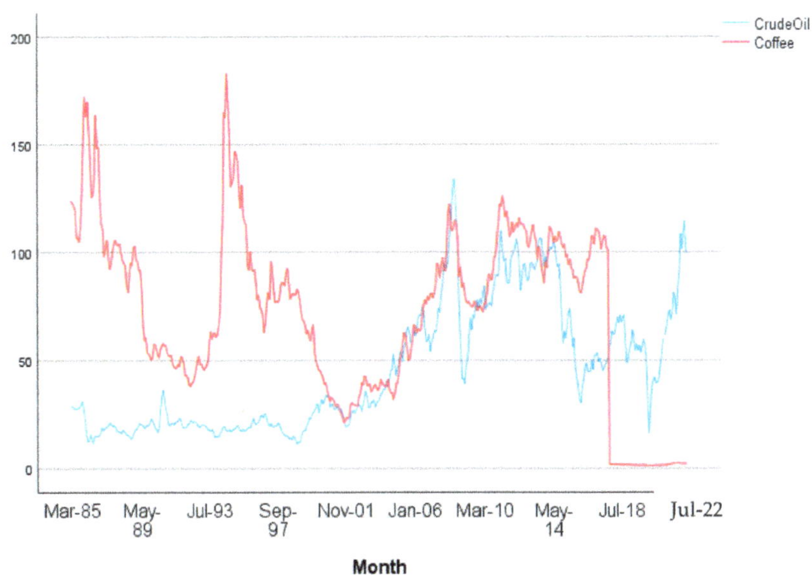

Figure 7. Time-Series Plot of Crude Oil and Coffee Price Patterns.

Proposition 8. *There is a positive relationship between oil and beef prices.*

The recent skyrocketing food prices have raised hyperinflation fears across the world. Coincidentally or not, since fast-rising food prices have accompanied the recent oil price hikes, many wonder if oil price has any bearing on food prices. In contrast with the finding of Onour [47], indicating no evidence of shared trends or cycles between oil and food (including beef) prices, I found a relatively strong correlation between oil and beef prices, with a Pearson correlation coefficient of 0.670. Figure 8 shows similar price trend patterns, even though oil prices look more volatile than beef prices. The regression analysis result with beef price as the dependent variable and crude oil price as the independent variable reaffirms that crude oil is a good predictor of beef price (with a standardized β coefficient of 0.670 and an adjusted R-square value of 0.447). However, I found that the oil and pork price relationship was not as strong as the oil and beef price relationship.

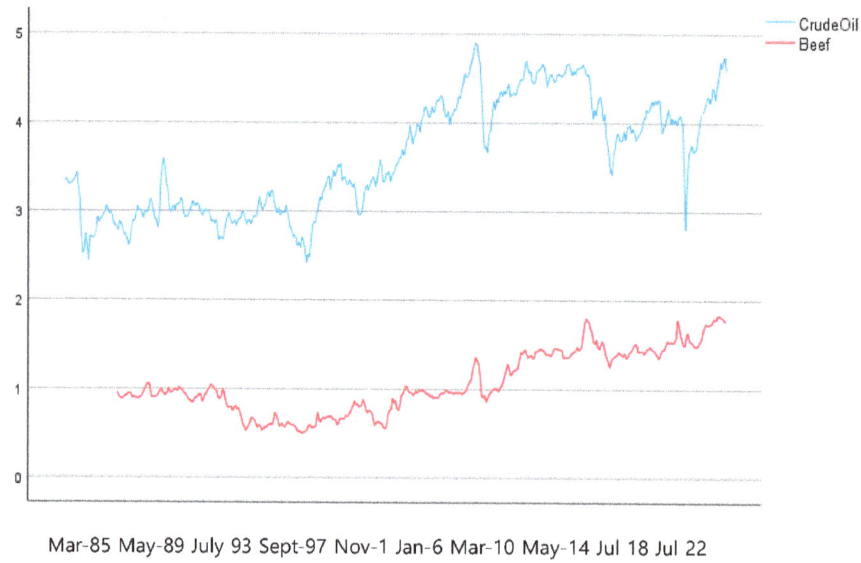

Figure 8. Time-Series (Logarithmic) Plot of Crude Oil and Beef Price Patterns.

5. Managerial Implications

So far, I have learned that crude oil price volatility tends to parallel the price volatility of many commodities, but not that of coffee. Primarily, I found that rising demand for alternative energy sources (especially biofuel) resulting from mounting oil prices created a new link between oil prices and the price volatility of agricultural commodities (especially corn) that can be transformed into biofuel materials. For a similar reason, I discovered that coal, as an alternative fossil fuel source, tends to have co-movement pricing patterns with oil. In a nutshell, oil price appears to have a profound impact on the prices of various commodities essential for everyday lives and industrial activities. Considering the critical role of oil in sustaining our standard of living, government policymakers and business decision makers should ensure the long-term stability of oil prices regardless of rapid environmental, social, economic, and geopolitical changes. Since such stability cannot be guaranteed without preventing or mitigating the risk of oil/gasoline supply chain disruptions, government policies and/or business strategies that can enhance resilience from supply chain disruptions should be developed. With that in mind, this paper also creates an oil/gasoline supply chain map that will allow political and business leaders to identify the vulnerability and potential bottlenecks of the oil/gasoline supply chain. Figure 9 graphically displays this map.

For example, if Russia's weaponization of its oil resources in response to the E.U.'s economic sanctions against Russia is the biggest culprit of worldwide oil supply shortages, a borderline between the upstream (refinery storage) and downstream (oil pipeline) levels of the oil supply chain is considered the most vulnerable chokepoint with the highest supply chain risk.

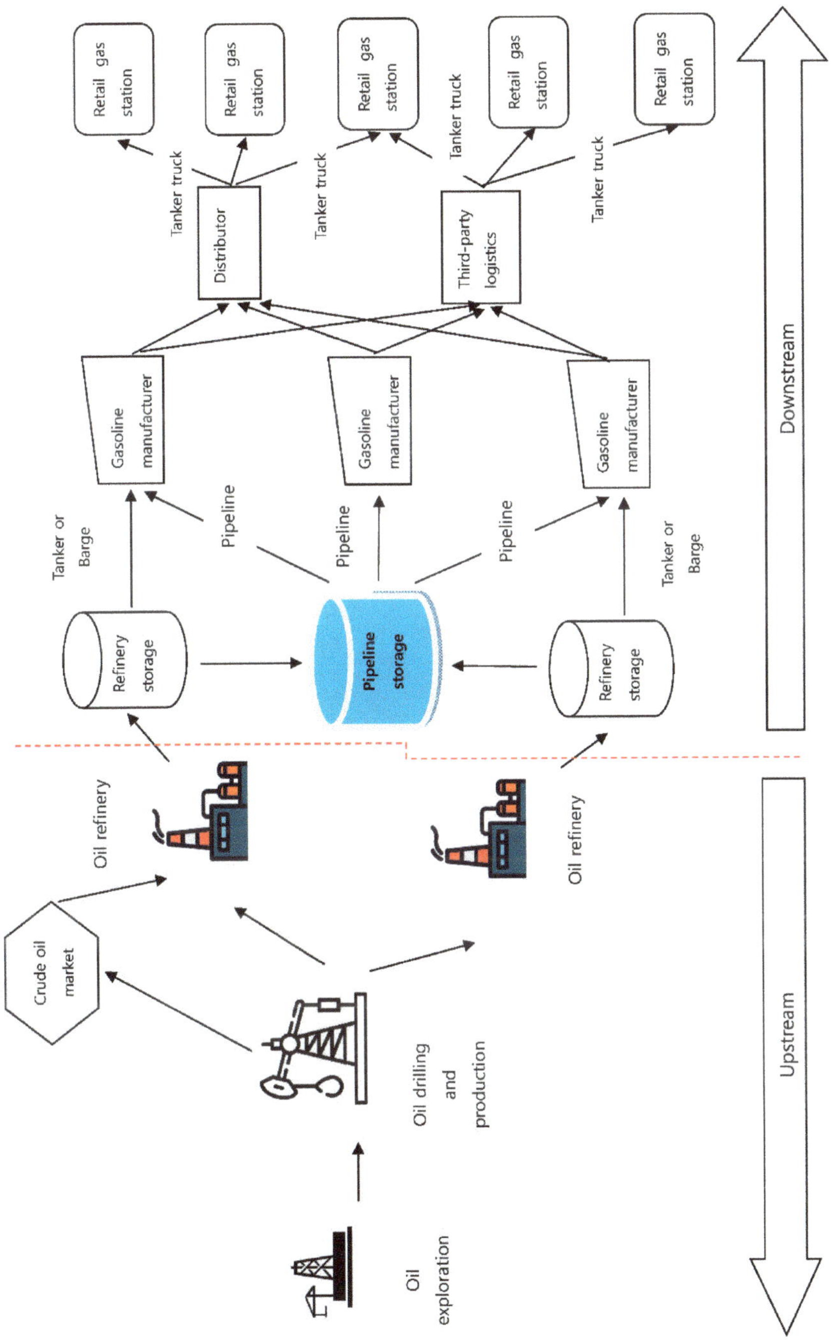

Figure 9. Oil/Gasoline Supply Chain.

6. A Summary and Concluding Remarks

In the era of worldwide energy and inflation crises created by the lingering COVID-19 pandemic, a series of natural disasters (e.g., flooding in Europe, East Asia, and the U.S.), and the prolonged war between Russia and Ukraine, the entire world has been swept into unprecedented economic turmoil. Volatile oil prices across the globe have further magnified this turmoil. This paper is one of the few attempts to investigate the functional relationship between the prices of crude oil and various types of commodities. In addition, this paper not only identifies a multitude of factors that can influence oil price and its volatility, but it also develops an oil/gasoline supply chain map that can visualize the weak points of an oil/gasoline supply chain vulnerable to disruptions. Through a series of experimental data analyses of four decades of primary energy sources (crude oil, coal, gasoline, and diesel) and commodities in high demand (corn, wheat, coffee, iron ore, aluminum, gold, silver, beef, and pork), I discovered the co-movement of crude oil and many commodities' pricing trend patterns. In particular, I found strong evidence indicating that oil price can be a good predictor of corn price, which, in turn, may influence food price. This finding implies that failure to stabilize oil price may substantially increase the cost of living and business expenditures. Although this paper did not present a clear causality between oil price and all the commodities, it reminds us of the crucial role of oil in sustaining our standard of living.

From a practical standpoint, this paper aids government policymakers and business executives in developing effective energy conservation policies and in strengthening the energy supply chain with enhanced resilience from various risks and uncertainties. In today's world, where many countries (developed or developing) are experiencing unprecedented energy crises and subsequent economic turmoil, establishing a more resilient energy supply chain helps humans better prepare for future energy crises. This paper also contributes to the existing body of energy literature by developing key propositions/hypotheses that raise future open research questions and by theorizing dynamic relationships between oil and daily necessities in the global commodity market. Despite these contributions, this paper is far from perfect in its current form. One of the major limitations of this paper includes a lack of scientific evidence indicating a clear causal relationship between crude oil prices and ongoing worldwide superinflation. As such, one of the fruitful areas of future research includes the examination of causal relationships between crude oil prices and consumer price indexes (CPIs) in both advanced and developing economies across the world. Another fruitful line of future research is the development of resilient energy supply chain strategies targeting specific countries or regions (e.g., E.U. nations) vulnerable to energy supply chain disruptions.

Funding: This research received no external funding.

Conflicts of Interest: The author declares no conflict of interest.

References

1. Statistica. Average Annual Brent Crude Oil Price from 1976 to 2022. 2022. Available online: https://www.statista.com/statistics/262860/uk-brent-crude-oil-price-changes-since-1976/ (accessed on 10 September 2022).
2. IEA. *World Energy Outlook*; Unpublished Report; IEA Energy Business Council: Paris, France, 2004.
3. IEO. International Energy Outlook. 2019. Available online: https://www.eia.gov/outlooks/ieo/ (accessed on 10 October 2019).
4. EIA. Annual Energy Outlook 2020. Available online: https://www.eia.gov/outlooks/aeo/ (accessed on 7 April 2020).
5. Golding, G. High Fuel Prices in the U.S. May Crimp Oil Demand Soon. 2022. Available online: https://www.dallasfed.org/research/economics/2022/0621 (accessed on 10 September 2022).
6. Asafu-Adjaye, J. The relationship between energy consumption, energy prices, and economic growth: Time series evidence from Asian developing countries. *Energy Econ.* **2000**, *22*, 615–625. [CrossRef]
7. Finn, M.G. Perfect competition and the effects of energy price increases on economic activity. *J. Money Credit. Bank.* **2000**, *32*, 400–416. [CrossRef]
8. Brown, S.P.; Yücel, M.K. Energy prices and aggregate economic activity: An interpretative survey. *Q. Rev. Econ. Financ.* **2022**, *42*, 193–208. [CrossRef]
9. Papapetrou, E. Oil price shocks, stock market, economic activity and employment in Greece. *Energy Econ.* **2001**, *23*, 511–532. [CrossRef]

10. Huang, B.; Hwang, M.; Peng, H. The asymmetry of the impact of oil price shocks on economic activities: An application of the multivariate threshold model. *Energy Econ.* **2005**, *27*, 455–476. [CrossRef]
11. Olomola, P.A. Oil price shock and aggregate economic activity in Nigeria. *Afr. Econ. Bus. Rev.* **2006**, *4*, 48–61.
12. Benkraiem, R.; Lahiani, A.; Miloudi, A.; Shahbaz, M. New insights into the U.S. stock market reactions to energy price shocks. *J. Int. Financ. Mark. Inst. Money* **2018**, *56*, 169–187. [CrossRef]
13. Balashova, S.; Serletis, A. Oil prices shocks and the Russian economy. *J. Econ. Asymmetries* **2020**, *21*, e00148. [CrossRef]
14. Carfora, A.; Pansini, R.V.; Scandurra, G. The causal relationship between energy consumption, energy prices and economic growth in Asian developing countries: A replication. *Energy Strategy Rev.* **2019**, *23*, 81–85. [CrossRef]
15. Dagoumas, A.S.; Polemis, M.L.; Soursou, S. Revisiting the impact of energy prices on economic growth: Lessons learned from the European Union. *Econ. Anal. Policy* **2020**, *66*, 85–95. [CrossRef]
16. Saghaian, S.H. The impact of the oil sector on commodity prices: Correlation or causation? *J. Agric. Appl. Econ.* **2010**, *42*, 477–485. [CrossRef]
17. Koirala, K.H.; Mishra, A.K.; D'Antoni, J.M.; Mehlhorn, J.E. Energy prices and agricultural commodity prices: Testing correlation using copulas method. *Energy* **2015**, *81*, 430–436. [CrossRef]
18. Fowowe, B. Do oil prices drive agricultural commodity prices? Evidence from South Africa. *Energy* **2016**, *104*, 149–157. [CrossRef]
19. López Cabrera, B.; Schulz, F. Volatility linkages between energy and agricultural commodity prices. *Energy Econ.* **2016**, *54*, 190–203. [CrossRef]
20. Wei Su, C.; Wang, X.; Tao, R.; Oana-Ramona, L. Do oil prices drive agricultural commodity prices? Further evidence in a global bio-energy context. *Energy* **2019**, *172*, 691–701. [CrossRef]
21. Volpe, M.A. Who Controls Oil and Gas Prices in the United States? *Business Week*. 31 May 2022. Available online: https://www.newsweek.com/who-controls-oil-gas-prices-united-states-1710982 (accessed on 20 September 2022).
22. McGraw-Hill. Crude Benchmark Analysis. 2006. Available online: https://web.archive.org/web/20060322115911/https://www.platts.com/Oil/Resources/News%20Features/crudeanalysis/index.xml (accessed on 20 September 2022).
23. Organization of the Petroleum Exporting Countries. OPEC Crude Oil Prices. 2022. Available online: https://data.nasdaq.com/data/OPEC/ORB-opec-crude-oil-price (accessed on 23 September 2022).
24. U.S. Energy Information Administration. What Drives Crude Oil Prices: Supply OPEC? 2022. Available online: https://www.eia.gov/finance/markets/crudeoil/supply-opec.php (accessed on 21 September 2022).
25. Barsky, R.B.; Kilian, L. Oil and the macroeconomy since the 1970s. *J. Econ. Perspect.* **2004**, *18*, 115–134. [CrossRef]
26. GEP. Russia-Ukraine War's Effects on the Oil and Gas Industry. 2022. Available online: https://www.gep.com/blog/mind/russia-ukraine-wars-effects-oil-and-gas-industry (accessed on 22 September 2022).
27. Teather, D. Oil Prices Close to Record as Katrina Shuts Down Production. *The Guardian*. 30 August 2005. Available online: https://www.theguardian.com/business/2005/aug/30/oilandpetrol.politics (accessed on 20 September 2022).
28. Chu, D.L. Japan Earthquake: Impact on Crude Oil, Fuel, and Nuclear Power. *Insider*. 14 March 2011. Available online: https://www.businessinsider.com/japan-earthquake-impact-on-crude-oil-fuel-and-nuclear-power-2011-3 (accessed on 22 September 2022).
29. Lade, G.E.; Lin, C.Y.C.; Smith, A. *Ex Post Costs and Renewable Identification Number (RIN) Prices under the Renewable Fuel Standard*; Unpublished Report, RFF-DP 15-22; Resources for Future: Washington, DC, USA, 2015.
30. Tayeb, Z. A Potential 'Black Swan' for U.S. Oil Prices Is Being Overlooked: Unreliable Electricity Grids. *Insider*. 3 July 2022. Available online: https://markets.businessinsider.com/news/commodities/oil-prices-energy-market-electricity-power-grid-black-swan-cornerstone-2022-7 (accessed on 22 September 2022).
31. International Energy Agency. Energy Statistics and Data Browser. 2022. Available online: https://www.iea.org/data-and-statistics (accessed on 24 September 2022).
32. Hult News. Renewable Energy Stays Strong Despite Plummeting Oil Prices. 2022. Available online: https://www.hult.edu/blog/renewable-energy-stays-strong-despite-plummeting-oil-prices/ (accessed on 11 September 2022).
33. Shively, B. How Much Primary Energy Is Wasted before Consumers See Value from Electricity. 2022. Available online: https://www.enerdynamics.com/Energy-Currents_Blog/How-Much-Primary-Energy-Is-Wasted-Before-Consumers-See-Value-from-Electricity.aspx (accessed on 25 September 2022).
34. EIA. What Countries Are the Top Producers and Consumers of Oil? 2022. Available online: https://www.eia.gov/tools/faqs/faq.php?id=709&t=6 (accessed on 15 September 2022).
35. Wisner, R. Corn, Ethanol and Crude Oil Prices Relationship—Implications for the Biofuels Industry. AgMRC Renewable Energy Newsletter. 2009. Available online: https://www.agmrc.org/renewable-energy/renewable-energy-climate-change-report/renewable-energy-climate-change-report/august-2009-newsletter/corn-ethanol-and-crude-oil-price-relationships---implications-for-the-biofuels-industry (accessed on 1 October 2022).
36. Trefis Team. Iron Ore & Crude Oil: The Similarities & Differences in the Market Dynamics of These Commodities. *Nasdaq*. 15 May 2017. Available online: https://www.nasdaq.com/articles/iron-ore-crude-oil-similarities-differences-market-dynamics-these-commodities-2017-03-15 (accessed on 18 September 2022).
37. Knoema. Aluminum Price Forecast: 2021, 2022 and Long Term to 2035. 2022. Available online: https://knoema.com/infographics/ffzioof/aluminum-price-forecast-2021-2022-and-long-term-to-2035 (accessed on 1 October 2022).

38. Kumar, V. Commodity Price Risk in 2022: Aluminum, Supply Chain & Demand Chain Executive, 2022. Available online: https://www.sdcexec.com/sourcing-procurement/sourcing-solutions/article/22018008/aranca-commodity-price-risk-in-2022-aluminum (accessed on 1 October 2022).
39. Leydesdorff, L.; Bensman, S. Classification and powerlaws: The logarithmic transformation. *J. Am. Soc. Inf. Sci. Technol.* **2006**, *57*, 1470–1486. [CrossRef]
40. Šimáková, J. Analysis of the relationship between oil and gold prices. *J. Financ.* **2011**, *51*, 651–662.
41. Bold Business Insights. Precious Insights–Do Oil Prices Determine Gold Prices? 2018. Available online: https://www.boldbusiness.com/energy/precious-insights-oil-prices-determine-gold-prices/ (accessed on 22 September 2022).
42. Baffes, J. Oil spills on other commodities. *Resour. Policy* **2007**, *32*, 126–134. [CrossRef]
43. Cartwright, P.A.; Riabko, N. Measuring the effect of oil prices on wheat futures prices. *Res. Int. Bus. Financ.* **2015**, *33*, 355–369. [CrossRef]
44. Reboredo, J.C. Do food and oil prices co-move? *Energy Policy* **2012**, *49*, 456–467. [CrossRef]
45. Maurice, N.E.; Davis, J. *Unraveling the Underlying Causes of Price Volatility in World Coffee and Cocoa Commodity Markets*; Unpublished Discussion Paper; UNCTAD: Geneva, Switzerland, 2011.
46. Vijayakumar, A.N. Impact of the U.S. dollar index and crude oil on prices of Arabica and Robusta varieties of Indian coffee. *Bus. Perspect. Res.* **2022**, 22785337221077399. [CrossRef]
47. Onour, I.A. Global food crisis and crude oil price changes: Do they share common cyclical features? *Int. J. Econ. Policy Emerg. Econ.* **2010**, *3*, 61–70. [CrossRef]

energies

MDPI

Article

Does New Digital Infrastructure Promote the Transformation of the Energy Structure? The Perspective of China's Energy Industry Chain

Lei Fan [1], Yunyun Zhang [2], Meilin Jin [3], Qiang Ma [4] and Jing Zhao [5,*]

1 School of Economics & Management, Nanjing Tech University, Nanjing 211816, China
2 Business School, Shandong Normal University, Jinan 250358, China
3 Institute of Food and Strategic Reserves, Nanjing University of Finance and Economics, Nanjing 210023, China
4 School of Economics and Management, Zhejiang University of Science and Technology, Hangzhou 310023, China
5 School of International Economics and Trade, Nanjing University of Finance & Economics, Nanjing 210023, China
* Correspondence: 9120201056@nufe.edu.cn

Abstract: In the context of carbon neutrality, the development of new digital infrastructure (NDI) and the improvement of digital capabilities are essential, in order to speed up the transformation of the energy structure. Based on the balanced panel data of 30 provinces in China from 2008 to 2019, we empirically analyzed the impact of NDI on the structural transformation of energy in China and its mechanisms of action. The results demonstrated that (1) NDI had a positive impact on China's energy transition, and the empirical results were robust. (2) The mediating effect showed that NDI had a positive impact on the transformation of energy structure, through improving green total factor productivity and green finance. (3) The heterogeneity analysis indicated that NDI made a more significant contribution to the transformation of the energy structure in regions with lower pollution levels and in those with energy cooperation policies. This study provides a policy reference for Chinese energy transition from the perspective of the digital economy.

Keywords: new digital infrastructure; transformation of the energy structure; energy industry chain

Citation: Fan, L.; Zhang, Y.; Jin, M.; Ma, Q.; Zhao, J. Does New Digital Infrastructure Promote the Transformation of the Energy Structure? The Perspective of China's Energy Industry Chain. *Energies* **2022**, *15*, 8784. https://doi.org/10.3390/en15238784

Academic Editors: Jiachao Peng, Le Wen, Jianzhong Xiao, Ming Yi and Mingyue (Selena) Sheng

Received: 26 October 2022
Accepted: 13 November 2022
Published: 22 November 2022

1. Introduction

To avoid the worst effects of climate change and to accelerate sustainable economic development, the world needs to phase out the use of non-clean energy sources [1]. The International Energy Agency (IEA) reported that global energy consumption increased by 50–75% between 1985 and 2020. Although the use of clean energy is increasing year by year, the proportion of fossil energy in primary energy consumption has been around 80% for nearly a century [2,3]. This high proportion of fossil fuel consumption is a barrier to the sustainable development of the global economy. The main driver of the current transition differs from the previous three, in that it is no longer the economic efficiency of new energy sources, but the response to climate change [4]. Resource depletion, the rising costs of non-clean energy sources, and technological innovation will further accelerate the energy transition process, making it imperative that the energy transition forms an important part of sustainable economic development [5–7].

As another important driving force to promote economic development and optimize economic structure, information and communications technology (ICT) has been widely recognized by scholars [8–10]. However, the new digital infrastructure (NDI) that supports cutting-edge ICT is a relatively neglected area of research. New digital infrastructure refers to the digital-based infrastructure formed by the new generation of information technologies, mainly including 5G, artificial intelligence, and the industrial Internet, and Internet of Things, which provide services that trigger fundamental changes in production patterns and economic structures [8,10]. The rate of change in the way NDI is supplied and the demand for digitally connected services in the last decade has been staggering. NDI has become a necessary factor of production for consumers to participate in advanced modern societies [11]. The services provided by NDI penetrate into the production and operation models of enterprises, improve the efficiency of industrial resource allocation, and optimize the regional industrial structure, as well as providing new ideas for solving the problem of energy transition [12–14].

The existing studies have concentrated on the relationship between the services provided by NDI and the energy transition, and the results are highly variable. Some scholars argue that the transformation of the energy structure with the services provided by NDI can accelerate the energy structure transition. Chung (2018) [15] suggested that the new energy systems with digital infrastructure services can accelerate the transformation of regional energy structure. Tang et al. (2013) [16] found that NDI helps to cultivate the green technology innovation capacity of firms and improves corporate governance. Other scholars have argued that it is uncertain the services provided by NDI contribute to the transformation of regional energy structure. Lange et al. (2020) [17] indicated that information and communication technologies increase global non-clean energy consumption by expanding energy demand. Shabani and Shahnazi (2019) [18] stated that the relationship between NDI services and energy consumption is not linear. The emergence of contradictory research conclusions is mainly related to the use of single indicators of NDI, the inconsistent construction standards of indicators, and a lack of heterogeneity in analysis. In addition, a large number of studies have found that both NDI and energy transition are related to green development, but few literature reports have explored whether there is a green development-related path between the two.

Based on provincial panel data in China, we focused on the impact of NDI on the energy transition, and explored the mechanisms and heterogeneity from a green development perspective. In fact, with the development of the digital economy, the new digital infrastructure is different from traditional infrastructure, which will have an important impact on regional green development. Therefore, it is necessary to identify the impact of new digital infrastructure on energy structure transformation. Our innovations include the following: (1) Previous studies examined the economic or environmental effects of traditional infrastructure, while ignoring the environmental effects of NDI, especially its impact on the energy structure. (2) We not only investigated the transmission mechanism of the impact of NDI on energy structure, but also investigated its heterogeneity in different regions, in order to investigate the impact of NDI on energy structure more comprehensively. (3) Under the dual backgrounds of the digital economy and carbon peak and carbon neutrality, it is of great practical significance to explore the impact of NDI on energy structure. These conclusions provide a theoretical basis for scientifically formulating new targeted digital basic energy service facilities.

The rest of the paper is organized as follows: Section 2 presents a mechanical analysis and the research hypotheses. Section 3 describes the data and empirical methodology. Section 4 discusses the empirical results. Section 5 summarizes the conclusions and policy implications.

2. Mechanism Analysis and Research Hypothesis

2.1. The Direct Effects

The efficient resource allocation advantages of NDI help to optimize energy allocation. ICT has a profound impact on energy management because it lowers costs and keeps systems up-to-date [19,20]. Cloud computing and big data analysis help to improve the efficiency of energy production, and wireless networks allow for timely optimization of energy allocation structures through online platforms [13]. The substitution and optimization effects of ICT on energy consumption contribute to the "computerization" of the production sector [21]. The above-mentioned optimization components, which rely on NDI, are more focused on the production and management of energy, optimizing the way energy is consumed by eliminating outdated capacities [22]. Efficient resource allocation is conducive to the transformation of regional energy structure. In view of the absolute advantages of varieties of energy and the comparative advantages of energy industries, the existing energy system requires a more timely and rapid energy allocation mechanism. For example, transmission network managers in Belgium are helping the network absorb more intermittent renewable energy by sharing computer platforms [23]. NDI provides a reliable technological path for building "smart cities" and promoting the coordination of low-carbon energy [24]. The market is also an important way to allocate resources. A sound market mechanism and flexible market design can both facilitate the energy transition. NDI increases the flexibility and timeliness of energy markets, thereby addressing some of the technical barriers faced in the development of regional energy structure transitions [25]. The high penetration of NDI can influence and even change group consumer behavior [13]. NDI changes the original method of information transmission and accelerates the transformation of the low-carbon behavior of energy consumers. For example, consumers can choose low-carbon technology application products and respond to their local government's call for low-carbon policies, and they can increase their awareness of green energy consumption [26]. Thongmak and Mathupayas (2016) [27] argued that effective dissemination of information helps consumers understand the current environmental situation and increases environmental empathy and environmental knowledge, thereby influencing their ecological consumption behavior. Existing empirical results also suggest that the development of ICT has contributed to renewable energy consumption. Moyer and Hughes (2012) [28] found that advances in communication technologies are generally positively associated with increased energy intensity and renewable energy generation. Zheng and Wang (2021) [13] found that a 1% increase in the level of mobile communication technology was associated with a 1.1% increase in renewable energy consumption in the short term and a 0.2% increase in the long term.

Based on the above analysis, the following hypothesis is proposed in this paper:

Hypothesis I: *NDI has a significant positive impact on the transformation of the regional energy structure.*

2.2. Indirect Impact

2.2.1. Green Production Level

NDI increases the green total factor productivity (GTFP) of a region. In terms of accelerating information flows, NDI facilitates the development of information technology, while the convergence of industrialization and modern technology contributes to the iteration and updating of green production technologies, making regional industrial production greener and more sustainable [6]. Yan et al. (2018) [29] demonstrated through empirical tests that trade in communication technologies can bridge the gap between developing economies. In terms of changing traditional production methods, Haftu (2019) [30] suggested the positive role of infrastructure in greening total factor productivity. Digital information technology can help build a more diverse labor supply, with human capital as a supporting condition; thus making labor demands adapt to green production methods. The diffusion of NDI services helps to spread the effects of green production, which in turn forces other firms in the same industry to improve their own GTFP.

The increase in level of green production has promoted the transformation of the energy structure. In terms of substitution effects, the energy transition is the substitution of one energy source for another in certain industries or sectors [31]. The improvement of the level of green production represents an improvement of renewable energy efficiency, and existing research also recognized the positive effect of green energy utilization technology on energy structure transformation in production activities [32]. In terms of the transformation of consumption, improved green production methods are a form of energy consumption efficiency improvement, and a sustainable energy structure built through green production promotes the development of green energy consumption preferences among consumers [33]. Quantitative changes in the level of green production are driving qualitative changes in energy transformation.

2.2.2. Green Finance Level

NDI promotes the development of regional green finance. At present, there is a lack of communication about the information mechanisms, in the development of green finance. NDI can distribute the environmental, social, and governance (ESG) information of enterprises to the public and increase the level of green financial support of high-quality enterprises. At the same time, enterprises will also conduct green reputation risk management, to ensure the financial support for their own green projects. The spillover effects of digital infrastructure further improve the green financial environment within the industry [34]. Not only that, but NDI can also help solve the problems of green finance regulation. Qing (2019) [35] suggested that the Indonesian government could make use of NDI to make up for the deficiencies of green information governance. Big data technologies provided by NDI can help commercial banks reduce unnecessary loan losses by optimizing energy-saving funds and green investment systems [36].

The development of green finance has promoted regional energy transformation. In terms of the function of cultivating dynamic energy, the development of green finance can help fill the huge investment gap in sustainable energy transition, provide sufficient funds for green technology innovation activities, and enhance the positive impact of innovation in renewable energy consumption [37,38]. In terms of the function of guiding resource allocation, green finance can guide social capital flow to green industries, improve the industrial structure, and promote the transformation of society, from a high-carbon economy, to a low-carbon economy [39]. Navarro (2019) [40] conducted a study on the feasibility of green finance, arguing that retail investors, producers, and financial institutions can promote the regional energy transition without compromising the interests of consumers, by creating green financial products.

Based on the above analysis, the following hypothesis is proposed in this paper:

Hypothesis II: *NDI further promotes regional energy structure transformation by enhancing GTFP and developing green finance.*

2.3. Heterogeneity Analysis

2.3.1. The Effect of NDI on Energy Structure Transformation Is Related to the Degree of Regional Pollution

In regions with different pollution levels, the role of NDI in the transformation of energy structure may be different, mainly due to environmental regulations and resource endowment [41,42]. Economic development is accompanied by high environmental costs and huge resource consumption, and green development has become the world consensus, which means that various regions need to reduce the environmental cost of economic development and maintain a balance between economic development and the ecological environment [43]. NDI provides equipment for ICT, social media, mobile technologies, and information networks, all of which contribute to information-based environmental governance [44]. Environmental regulation further affects the proportion of renewable energy used, through "compliance cost effects" and "innovation offset effects" [16,45,46]. Therefore, we speculate that NDI can improve the level of regional environmental supervision, thereby reducing regional pollution. The transformation of the energy structure is one of the manifestations of the improvement of regional pollution. Natural resource endowments will also affect regional energy transformation plans. The industrial structure being formed by relying on the regional natural resource endowment has led to the heterogeneity of air pollution among provinces. The degree of pollution of regions rich in non-renewable resources is higher, and the energy transformation is more difficult [47–49]. Not only that, regional economic conditions also strongly influence the energy transition plans of local governments, and economic factors affect the transition from fossil fuels or non-clean energy, to renewable or clean energy [50]. The coverage rate of NDI in economically developed regions is high, and the phenomenon of the resource curse is common in regions with high natural resource endowments [51,52]. Therefore, it is speculated that in regions with abundant natural resources, the positive effect of NDI on the energy transition is weak.

2.3.2. The Effect of NDI on Energy Structure Transformation Is Related to Regional Energy Cooperation Policies (ECP)

ECP can help strengthen the positive role of NDI in the transformation of regional energy structures. In terms of the content of the ECP, the ECP conforms to the latest energy production trends and improves the existing energy structure through multilateral cooperation [53,54]. For energy structure transformation, the energy production cooperation, energy investment cooperation, and energy infrastructure connection content in the ECP provide a larger development platform for NDI [6,55]. In terms of the drivers of ECP, there are two main charagories of ECP implementation region: First, the region has sufficient demand to reduce the dependence on non-renewable energy. Second, the region has relatively good energy cooperation conditions. These two conditions are conducive to the transformation of the energy structure [56]. Economic, environmental, and political factors help drive the success of energy cooperation [57]. Therefore, policymakers will delineate ECP areas with reference to the driving factors of energy cooperation success, which reflects the economic, environmental, and political differences between areas covered by ECP and non-covered areas. According to the previous analysis, the differentiated effect of NDI on the transformation of the energy structure can be explained. In summary, the mechanism diagram of this paper is shown in Figure 1.

Based on the above analysis, the following hypothesis is proposed in this paper:

Hypothesis III: *From the perspective of heterogeneity, in areas with low levels of environmental pollution and under energy cooperation policies (ECP), the positive effect is more obvious.*

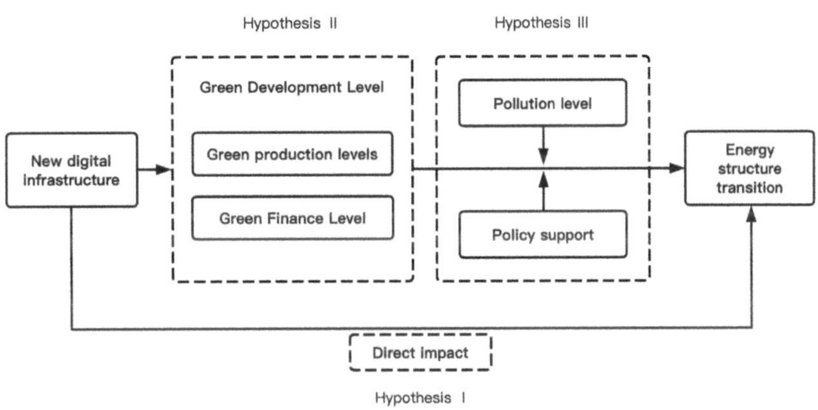

Figure 1. The impact mechanism of the NDI on energy structure transition.

3. Method

3.1. Model

To verify the impact of NDI on energy transition, we chose an OLS model with multi-dimensional panel fixed effects to mitigate the bias of results, by controlling for multi-dimensional individual effects. We address the problem of missing variables with individuals and time, by controlling for the individual effects and annual effects of provinces.

This paper constructs the following model:

$$\text{energy}_{it} = \beta_0 + \beta_1 \text{infra}_{it} + \gamma \sum \text{control}_{it} + \lambda_i + \mu_t + \varepsilon_{it} \quad (1)$$

where energy_{it} is the energy consumption structure of province i in year t, infra_{it} is the NDI of province i in year t, and control_{it} is the set of control variables, specifically government self-sufficiency (suf_{it}), level of environmental regulation (er_{it}), level of urbanization (urban_{it}) and its squared term (urban_{it}^2), economic development level (pgdp_{it}), and level of education (egdp_{it}). λ_i is the province fixed effect, μ_t is the time fixed effect, and ε_{it} are random errors.

3.2. *Variables*

3.2.1. Explained Variable: Energy Structure (energy_{it})

Referring to the method of Adebayo et al. (2021) [58], the ratio of coal consumption to total energy consumption is used to measure the energy structure of each province. Coal consumption and total energy consumption were converted into standard coal, with reference to the China Energy Statistical Yearbook. The conversion coefficients of major energy categories are shown in Table 1.

Table 1. Reduction coefficients of major energy categories in the energy industry chain.

Energy Category	Reduction Factor	Energy Category	Reduction Factor
Raw coal	0.7143	Gasoline	1.4714
Washed coal	0.9000	Diesel	1.4571
Coal products	0.5286	Kerosene	1.4714
Coke	0.9714	Fuel oil	1.4286
Coke oven gas	5.7140	Liquefied petroleum gas	1.7143
Natural gas	13.3000	Electricity	1.2290
Liquefied natural gas	1.7572	Thermal	0.0341
Crude oil	1.4286	Others	1.000

3.2.2. Explanatory Variables: NDI (infra_{it})

Referring to the practice of Zhao (2022) [59], a four-dimensional index system of long-distance optical cable lines, mobile phone switch capacity, industrial robot installation, and the number of Internet access ports is used, and the entropy weight method is selected, to determine the weight of each index, and, finally, the new digital number is calculated. Infrastructure metrics:

3.2.3. Control Variables

(1) Government self-sufficiency rate (suf_{it}). Fiscal self-sufficiency is a significant criterion for judging whether the development of a regional government is healthy. Local governments can use public finance to solve the problem of social and economic inequality, so as to promote the low-carbon transformation of the region [60]. Saygin et al. (2015) [61] suggested that a government can push the transformation of the energy structure in a region by providing guiding policies to develop technological innovation in renewable energy. Referring to the practice of Yan et al. (2022) [62], we adopt the ratio of the revenue in the general budget of the local government to the expenditure in the general budget of the local government to represent the government's self-sufficiency rate.

(2) The intensity of environmental regulation (er_{it}). When the government implements a series of environmental regulation policies, polluters will predict increase in the intensity of environmental regulation in the future, so as to strengthen their current utilization of such energy, or force enterprises to adopt clean energy and advanced energy-saving and emission-reduction technologies, by improving industry standards [63]. Referring to the practice of Peng et al. (2020) [64], we select the com-

prehensive utilization rate of solid waste, to measure the intensity of environmental regulation in each province.

(3) The level of urbanization ($urban_{it}$). At a particular stage in economic development, the increase of energy consumption follows an "inverted U" curve with rising urbanization levels, which leads to population clustering, changes in energy consumption patterns, and technological innovation. These changes push the structure of energy consumption toward optimization. Referring to Liu et al. (2022) [65], this article uses the proportion of urban population to characterize the urbanization level of a region and introduces a squared term of the urbanization level ($urban_{it}^2$), to ensure the adequacy of urbanization level in explaining the energy consumption structure.

(4) The level of economic development ($pgdp_{it}$). Taghizadeh and Rasoulinezhad (2020) [66] stated that there is a positive correlation between economic development and regional energy transition. We refer to the practice of Acheampong et al. (2021) [67], which used the logarithm of gross domestic product (GDP) per capital to measure the level of regional economic development.

(5) Educational level ($egdp_{it}$). Level of education may influence the environmental awareness of residents, which in turn affects their acceptance and support for the energy transition Tang et al. (2013) [16] also believed that educational level also played a certain role in the process of energy structure optimization. We reference Li et al. (2022) [68] and uses the share of local fiscal expenditure on education in regional GDP to measure the level of education. To sum up, the specific variable description is shown in Table 2.

Table 2. Variable definitions.

Variable Classification	Variable	Definition
Explained variable	Energy structure	Proportion of energy consumed by coal compared to total energy consumed
Explanatory variable	Infra	New digital infrastructure
	Suf	Local government revenue to expenditure ratio
	Er	Comprehensive utilization rate of solid waste
Control variables	Urban	Level of urbanization
	$Urban^2$	Square of the level of urbanization
	Pgdp	Logarithm of GDP per capital
	Egdp	Education expenditure as a percentage of GDP
Intermediate variables	Gtfp	Green total factor productivity
	Gfin	Green finance index

3.3. Data Description

This paper takes the data of 30 provinces, autonomous regions, and municipalities in China (excluding Hong Kong, Macao, Taiwan, and Tibet) as the research object. The time span of the data is from 2008 to 2019. The variable data in this article were mainly obtained from the "China Energy Statistical Yearbook", "China Statistical Yearbook", "China Industrial Statistical Yearbook", "China Urban Statistical Yearbook", "China Industrial Economic Statistical Yearbook", "China Environmental Statistical Yearbook", and "China Insurance Yearbook and Statistical Yearbooks of various provinces". Industrial robot data were obtained from the International Federation of Robotics. Table 3 shows a descriptive statistical analysis of the above variables.

Table 3. Descriptive statistics of variables.

Var	N	Mean	Std	Min	Q25	Q50	Q75	Max
energy	360	0.414	0.152	0.012	0.312	0.438	0.519	0.724
infra	360	0.194	0.147	0.012	0.088	0.150	0.251	0.941
suf	360	0.538	0.194	0.189	0.400	0.487	0.706	0.960
er	360	0.665	0.191	0.254	0.511	0.644	0.834	0.998
urban	360	0.559	0.131	0.291	0.469	0.543	0.616	0.942
pgdp	360	1.629	0.419	0.680	1.352	1.594	1.870	2.860
egdp	360	0.040	0.015	0.019	0.029	0.037	0.048	0.112

Figures 2 and 3 plot the average level of China's energy consumption structure and NDI from 2008 to 2019. Specifically, the proportion of coal consumption in the central and western regions is relatively high, while the proportion of coal consumption in the eastern region is relatively low. The level of NDI in the central and western regions is generally low, and that in the eastern regions is generally high. Regions with a high level of NDI are characterized by a developed economy, the concentration of scientific and technological talent, and a low endowment of non-clean energy resources. Regions with a high proportion of coal energy consumption have the characteristics of an underdeveloped economy, large loss of scientific and technological talents, and high endowment of non-clean energy resources.

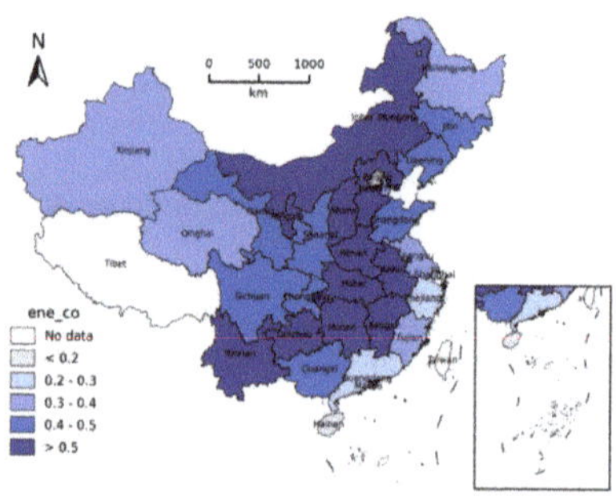

Figure 2. Distribution of energy consumption structure (2008–2019).

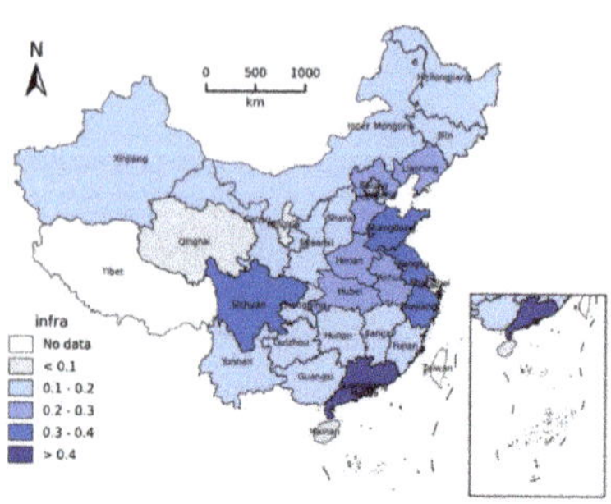

Figure 3. Distribution of new digital infrastructure (2008–2019).

4. Results

4.1. Benchmarking

We used a linear regression model to test the direct impact of NDI. Table 4 shows the direct impact of NDI on the energy transition. Columns (1) and (2) do not include control variables, while columns (3) and (4) include control variables. Columns (1) and (3) fixed the year, while columns (2) and (4) fixed the year and province. From the results in column

(4) of Table 4, the impact of NDI on the energy structure remained significant at 1% and was negative. The impact coefficient was −0.234, which indicates that NDI is conducive to transformation of the energy structure.

Table 4. Direct impact of NDI on energy structure transition.

Variables	(1) Energy	(2) Energy	(3) Energy	(4) Energy
infra	−0.126 *	−0.179 ***	−0.118 *	−0.196 ***
	(−1.67)	(−4.27)	(−1.72)	(−4.47)
suf			−0.183 ***	−0.064
			(−3.66)	(−1.26)
er			−0.075 *	−0.047 *
			(−1.87)	(−1.67)
urban			−1.119 ***	−0.538 *
			(−3.26)	(−1.90)
$urban^2$			0.689 ***	0.257
			(2.62)	(1.08)
pgdp			−0.206 ***	−0.102 **
			(-4.71)	(−2.43)
egdp			−5.312 ***	−0.702
			(−8.77)	(−1.13)
Constant	0.439 ***	0.449 ***	1.533 ***	0.929 ***
	(26.61)	(53.57)	(11.98)	(9.55)
Year Fixed Effect	Yes	Yes	Yes	Yes
Province Fixed Effect	No	Yes	No	Yes
Observations	360	360	360	360
R-squared	0.115	0.943	0.578	0.948

Notes: ***, **, and * represent the significance levels of 1%, 5%, and 10%, respectively, with the same below.

To ensure the robustness of the regression results, we used a replacement regression method for the test. We used the SYS-GMM model to estimate the impact of NDI on the regional energy transition. The p-value of AR(1) in Table 4 is less than 0.05, and the p-value of AR(2) is greater than 0.1, which means that the hypothesis that the remaining terms have first-order auto-correlation was accepted, and the hypothesis that the remaining terms have second-order auto-correlation was rejected. The p-value of the Sargan test was greater than 0.05, which confirmed the validity of the instrumental variables. The above results indicate that the choice of variables was valid and that the model is appropriate. Table 4 shows the regression results of the SYS-GMM model. The results show that infra had a significant negative impact on energy, which indicates that NDI had a catalytic effect on regional energy transition.

In addition, considering the endogenous of the article, we used the instrumental variable to test the model. Referring to Oughton (2021) [69], we used Internet broadband access ports (intacc) as an instrumental variable for NDI, which was primarily obtained from the National Bureau of Statistics of China. The number of Internet broadband access ports reflects the penetration rate of NDI, so we chose the number of Internet broadband access ports as an instrumental variable for NDI. Internet broadband access ports do not directly influence the transformation of the regional energy structure, and NDI is needed to provide some support, so the number of Internet broadband access ports as an instrumental variable satisfies the exclusivity requirement. According to column (3) of Table 5, the instrumental variable passed the F test, which means that the increased Internet broadband access ports improved the regional energy structure, which is consistent with the results in Table 4, demonstrating the reliability of the empirical results.

Table 5. Robustness test of NDI for regional energy transition.

Variables	(1) SYS-GMM Energy	(2) SYS-GMM Energy	(3) IV Energy
L.energy	0.946 ***	0.905 ***	
	(45.87)	(30.68)	
infra	−0.042 ***	−0.051 **	−0.166 ***
	(−2.98)	(-2.38)	(−2.96)
suf		−0.032	−0.067
		(−1.64)	(−1.02)
er		−0.007	−0.003
		(−0.38)	(−0.09)
urban		0.056	−0.713 *
		(0.32)	(−1.74)
urban^2		−0.020	0.362
		(−0.15)	(1.06)
pgdp		-0.013	−0.106 **
		(−1.01)	(−2.00)
egdp		−0.417 *	−0.041
		(−1.91)	(−0.06)
Constant	0.018 *	0.074	
	(1.68)	(1.28)	
Year Fixed Effect	Yes	Yes	Yes
Province Fixed Effect	Yes	Yes	Yes
Observations	330	330	270
Number of Province	30	30	
Ar1 (*p* value)	0	0	
Ar2 (*p* value)	0.905	0.751	
Sargan (*p* value)	0.288	0.227	
KPLM			191.2
CDWaldF			545.9

Note: The standard errors in brackets, ***, **, *, indicate significant at 1%, 5% and 10% respectively, the same below.

4.2. Mechanism Inspection

Referring to the method of Shen et al. (2021) [70], this paper uses a mediation effect model to study the impact mechanism of NDI on energy transition.

$$\text{gtfp}_{\text{it}} = \alpha_1 \text{infra}_{\text{it}} + \gamma \sum \text{control}_{\text{it}} + \lambda_{\text{i}} + \mu_{\text{t}} + \varepsilon_{\text{it}} \quad (2)$$

$$\text{energy}_{\text{it}} = \alpha_2 \text{infra}_{\text{it}} + \alpha_3 \text{gtfp}_{\text{it}} + \gamma \sum \text{control}_{\text{it}} + \lambda_{\text{i}} + \mu_{\text{t}} + \varepsilon_{\text{it}} \quad (3)$$

$$\text{gfin}_{\text{it}} = \alpha_4 \text{infra}_{\text{it}} + \gamma \sum \text{control}_{\text{it}} + \lambda_{\text{i}} + \mu_{\text{t}} + \varepsilon_{\text{it}} \quad (4)$$

$$\text{energy}_{\text{it}} = \alpha_5 \text{infra}_{\text{it}} + \alpha_6 \text{gfin}_{\text{it}} + \gamma \sum \text{control}_{\text{it}} + \lambda_{\text{i}} + \mu_{\text{t}} + \varepsilon_{\text{it}} \quad (5)$$

Formulas (2) and (3) verify the mediation effect of GTFP (gtfp), and the effect value of the mediation effect in the total effect is $(\alpha_1\alpha_3)/\beta_1$; Formulas (4) and (5) verify the mediation effect of green finance index (gfin), where the effect value of the mediator effect over the total effect is $(\alpha_4\alpha_6)/\beta_1$.

Columns (1) and (2) of Table 6 use GTFP as an intermediary variable, column (1) shows that infra had a significant positive impact on GTFP, which means that NDI improved the GTFP. Column (2) shows that both infra and GTFP were negatively correlated with energy, which means that NDI not only directly contributed to the transformation of the energy structure, but also improved the energy structure by promoting green production. Therefore, GTFP plays an intermediary role in the transformation of energy structure and NDI. The mediating effect accounted for 23.1% of the effective value of the total effect.

Table 6. Mechanism test.

Variables	(1) Gtfp	(2) Energy	(3) Gfin	(4) Energy
gtfp		−0.024 **		
		(−2.42)		
gfin				−0.390 ***
				(−4.12)
infra	1.886 ***	−0.169 ***	0.054 **	−0.175 ***
	(5.85)	(−2.99)	(2.10)	(−4.07)
suf	0.276	−0.054	0.037	−0.050
	(0.88)	(−1.04)	(1.24)	(−1.00)
er	0.617 ***	−0.033	0.044 ***	−0.030
	(3.46)	(−1.09)	(2.65)	(−1.08)
urban	−15.396 ***	−0.793 **	−1.480 ***	−1.116 ***
	(−8.34)	(−2.33)	(−8.94)	(−3.60)
$urban^2$	12.815 ***	0.512 *	0.663 ***	0.516 **
	(8.31)	(1.80)	(4.79)	(2.15)
pgdp	1.184 ***	−0.064	0.093 ***	−0.066
	(4.32)	(−1.37)	(3.78)	(−1.57)
egdp	2.431	−0.724	0.530	−0.495
	(0.64)	(−1.15)	(1.46)	(−0.81)
Constant	3.036 ***	0.943 ***	0.546 ***	1.142 ***
	(4.84)	(8.75)	(9.63)	(10.57)
Year Fixed Effect	Yes	Yes	Yes	Yes
Province Fixed Effect	Yes	Yes	Yes	Yes
Observations	330	330	360	360
R-squared	0.944	0.950	0.961	0.950

Notes: ***, **, and * represent the significance levels of 1%, 5%, and 10%, respectively, with the same below.

Columns (3) and (4) of the model results in Table 5 use the green finance index as an intermediary variable. Column (3) shows that infra had a significant positive effect on gfin, implying that NDI increased the level of green finance. Column (4) shows that both infra and gfin were significantly negatively related to energy, indicating that green finance played a significant intermediary role in the energy structure transition, with the intermediary effect accounting for approximately 10.7% of the total effect. This result implies that NDI improved the regional energy structure by increasing the penetration of green finance. This finding is similar to the conclusion of Wang et al. (2021) [71] that green finance development contributes to the shift from conventional to renewable energy consumption.

4.3. Heterogeneity Analysis

In fact, Figure 2 demonstrates that the level of NDI varies by region. Differences in their energy endowments and environmental monitoring lead to differences in the energy structure across countries. As a result, the impact of NDI on the transformation of the energy structure may vary by region.

4.3.1. The Impact of NDI on the Energy Transition in Provinces with Different Pollution Levels

Table 7 shows the impact of NDI on the energy transition in regions with various levels of pollution. Columns (1) and (2) are the comprehensive pollution scores calculated using the entropy weighting method, Columns (3) and (4) are the comprehensive pollution scores calculated using the principal component method. Specifically, the inhibitory effect of infra on energy passes the test for less polluted areas (pollution index below the median). In less polluted areas, NDI promoted the regional energy structure transformation. For areas with high pollution levels, the effect of infra on energy did not pass the test. NDI had a greater impact on the energy transition in regions with a lower pollution level. The penetration of NDI could help reduce the dependence on regional resource endowments, improve the ability of environmental supervision, and promote the transformation of the energy structure.

Table 7. The impact of NDI on energy transition in provinces with different pollution levels.

Variables	(1) Poent ≤ Med	(2) Poent > Med	(3) Popri ≤ Med	(4) Popri > Med
infra	−0.538 ***	−0.090	−0.552 ***	−0.100
	(−6.08)	(−1.28)	(−6.10)	(−1.52)
suf	0.016	−0.082	−0.028	−0.091
	(0.20)	(−1.29)	(−0.35)	(−1.44)
er	−0.181 ***	0.093 **	−0.160 ***	0.056
	(−4.98)	(2.17)	(-4.31)	(1.35)
urban	0.357	−1.879 ***	0.109	−1.921 ***
	(0.85)	(-3.07)	(0.27)	(−4.03)
$urban^2$	−0.201	1.605 ***	−0.123	1.574 ***
	(−0.64)	(2.82)	(-0.41)	(3.48)
pgdp	−0.130 ***	−0.000	−0.221 ***	0.014
	(−2.87)	(−0.00)	(−3.84)	(0.24)
egdp	−1.487 **	−0.483	−1.475 **	−0.701
	(−2.22)	(−0.33)	(−2.14)	(−0.55)
Constant	0.666 ***	1.013 ***	0.979 ***	1.060 ***
	(5.34)	(6.97)	(6.37)	(7.41)
Year Fixed Effect	Yes	Yes	Yes	Yes
Province Fixed Effect	Yes	Yes	Yes	Yes
Observations	180	178	179	180
R-squared	0.960	0.949	0.959	0.955

Notes: *** and ** represent the significance levels of 1%, 5%, and 10%, respectively, with the same below.

There are two possible reasons why NDI promoted the energy transition in less polluted provinces. The first reason is the resource endowments of the regions. Areas with low pollution levels are usually not rich in resource endowments of non-clean energy, and their industrial development is less dependent on such energy sources. NDI has a high penetration, which provides a stronger impetus for low-carbon energy transformation in low-pollution areas [72]. The second reason is the pollution regulation in the region. Environmental supervision policies are regarded as one of the important ways to reduce environmental pollution. The tighter the monitoring of pollution, the lower the level of environmental pollution [73]. Therefore, when the local government pays attention to environmental pollution, NDI has a more significant effect on improving the government's pollution supervision ability.

4.3.2. The Impact of NDI on Energy Transition in Energy Cooperation Regions

Table 8 shows the impact of NDI on regional energy transformation under the ECP. We selected the "Belt and Road" policy as the representative of ECP [74–77]. Specifically, in the policy implementation areas of energy cooperation, the effect of infra on energy passed the 1% significance test. The results indicate that NDI had a greater impact on the energy transition in areas with the implementation of an energy cooperation policy. In contrast, in non-energy policy implementation areas, the effect of infra on energy did not pass the significance test. This result suggests that the impact of NDI on the energy transition in the non-energy cooperation policy implementation areas was not significant. The empirical tests found that NDI had a greater impact on the energy transition in energy cooperation policy regions, while the impact was not significant in non-energy cooperation policy regions. The reasons for this may be related to geographical location, and the government needs to consider geographical location and regional trade networks of China's trading partners when selecting provinces to implement the One Belt, One Road policy. Therefore, in regions with existing trade bases, the information supervision, information sharing, and information exchange services provided by NDI enhanced the efficiency and stability of energy cooperation, thereby accelerating the pace of regional energy transformation.

Table 8. Impact of NDI on energy transition in regions with or without in energy cooperation.

Variables	(1) Road = 0	(2) Road = 1	(3) Road = 0	(4) Road = 1
infra	0.006	−0.156 ***	0.024	−0.220 ***
	(0.08)	(−3.23)	(0.33)	(−4.01)
suf			−0.046	0.020
			(−0.67)	(0.29)
er			0.092 **	−0.092 ***
			(1.99)	(-2.77)
urban			0.004	−1.170 **
			(0.01)	(−2.36)
$urban^2$			0.350	0.846 *
			(1.29)	(1.87)
pgdp			−0.014	−0.048
			(−0.21)	(−0.90)
egdp			0.811	−0.545
			(0.63)	(−0.83)
Constant	0.448 ***	0.416 ***	0.277 *	0.952 ***
	(29.78)	(45.93)	(1.88)	(6.39)
Year Fixed Effect	Yes	Yes	Yes	Yes
Province Fixed Effect	Yes	Yes	Yes	Yes
Observations	156	204	156	204
R-squared	0.954	0.950	0.958	0.957

Notes: ***, **, and * represent the significance levels of 1%, 5%, and 10%, respectively.

5. Discussion

In the following, we present the main results of the effects mentioned above, pointing out the impact of NDI on energy structure transition. In Section 4, we test the hypotheses presented in this paper using empirical models and analyze the empirical results, the findings of which are summarized in Table 9. This paper discusses the effect, transmission mechanism, and conditions of NDI on energy structure transformation. The research showed that NDI had a significant promotional effect on energy structure transformation, and GTFP and green finance played an important role. In fact, China's energy industry chain is characterized by many links and long chains. NDI will help improve the high-end link capacity of the industrial, chain as well as the autonomy and controllability, so as to improve the functioning and replace the traditional energy system; accelerate the construction of a clean, low-carbon, safe, and efficient energy system; and facilitate the construction of a modern energy system. In addition, green finance and GTFP will accelerate the transformation of China's energy structure. Green finance provides long-term and low-cost financial support for the transformation of energy structure. Improving GTFP is the core of improving energy efficiency and promoting the transformation of energy structure. The conclusion of this paper shows that, in the context of carbon peaking and carbon neutrality, it is necessary to give full play to the importance of NDI in promoting the transformation of China's energy structure.

Table 9. Summary of results: three impacts of NDI on energy transition.

	Empirical Main Findings	Explanation
Effect I: Direct effect	NDI has a significant positive impact on the transformation of regional energy structure.	① NDI optimizes energy management systems and improves energy allocation efficiency. ② NDI increases the flexibility and timeliness of the energy market. ③ NDI enhances awareness of green energy consumption and accelerates the shift in low-carbon consumer behavior.
Effect II: Intermediary effects	NDI contributes to the transformation of regional energy structure by increasing GTFP.	① NDI facilitates the diffusion of green production technologies into industry, thereby increasing the efficiency of renewable energy use. ② NDI builds a diverse labor supply system, to meet the labor demands of green energy production methods. ③ NDI speeds up the flow of information and pushes companies toward green production, by creating a preference for green energy consumption.
	NDI promotes the transformation of regional energy structure through the development of green finance.	① NDI improves green information communication mechanisms, increases green financial support for quality companies, and fills the investment gap in the energy transition. ② NDI can optimize green investment systems and guide social capital into regional energy transition.
Effect III: Moderating effects	The positive effect of NDI on the transformation of the energy structure is evident in areas with low levels of environmental pollution.	① In areas of strong environmental governance awareness, NDI enhances environmental regulation and influences the transformation of the energy structure through the "cost of compliance effect" and "innovation offset effect". ② The positive effect of NDI on the energy transition is undermined by a high level of industrial energy structure dependence in areas with a large resource endowment in non-renewable energy sources and by the underdevelopment of regional economies.
	The positive effect of NDI on the transformation of the energy structure is evident in regions adopting the Energy Cooperation Policy (ECP).	① ECP regions generally have strong energy transition aspirations, raising the positive role of NDI for the energy transition. ② The geographical advantages and industrial needs of the regions involved in the ECP provide favorable conditions for energy cooperation and strengthen the positive role of NDI.

6. Conclusions and Policy Implications

Under the "dual carbon" goal, the rapid development of NDI construction has a profound impact on regional energy transformation. Based on provincial panel data in China, this paper explored the impact of NDI on regional energy structure and its mechanisms of action, from the perspective of green production and green finance. The main findings were as follows: NDI has a direct and significant impact on regional energy transition, and NDI facilitates regional energy transition. NDI not only directly affects the energy transformation of regions, but also has an indirect impact on regional energy transformation through GTFP and green financial. The intermediary effect of GTFP was 23.1%, and the intermediary effect of green finance was 10.7%. This conclusion provides a clearer explanation for the potential green mechanism of NDI and energy transition, and provides new ideas for improving regional energy structure. NDI has different impacts on pollution levels and energy transition policies in different regions. NDI has a significant positive effect on the energy transition in areas with low pollution levels or ECP, while it does not have a significant effect on the energy transition in areas with high pollution levels or without ECP policies. This means that the impact of NDI on energy transition is prominent in regions where the resource endowment is not abundant and the environmental supervision awareness is strong. These findings respond to, and expand on, the current debate on the relationship between NDI and the energy transition.

Based on the above findings, we can make the following policy recommendations:

(1) The government should pay attention to the construction of NDI and give full play to the positive role of NDI in regional energy transformation. Specifically, it should

follow the trend under the "dual carbon" goal; the rapid development of NDI has a profound impact on energy transformation. In addition, from the perspective of managerial implications, on the one hand, enterprises can enhance their green technology innovation ability by increasing R&D investment, such as promoting the technological innovation of renewable energy, including water energy, wind energy, solar energy, and tidal energy, so as to realize the transformation and upgrading of energy structure. On the other hand, the government needs to explore the mechanisms and practice of carbon reduction, with the demand side as the driving force to improve energy efficiency, and enhance the internal driving force of energy structure transformation.

(2) Governments should implement targeted energy transition strategies using the impact of NDI on energy transition. First of all, it is necessary to develop GTFP in industry, promote the updating and integration of modern technology and green production technology, and increase the utilization of renewable energy. In addition, we need to develop the level of regional green finance through NDI, compensate for the lack of green financial regulation, raise the efficiency of green investment, and provide new talent for regional energy transition.

(3) According to their level of new digital technology facilities and energy base, different regions should implement targeted energy policies. Specifically, areas with low pollution levels should strengthen the construction of NDI, give full play to its advantages, upgrade their own pollution supervision systems, and provide a model for optimizing their energy structure. Areas with high pollution levels should eliminate their excessive dependence on non-clean energy resources as soon as possible, raise awareness of pollution control, and actively introduce advanced technologies of renewable energy, so as to optimize the regional industrial energy structure. In addition, it is important to continue international energy cooperation, to learn from the successful experiences of cooperation, to introduce advanced renewable energy application technologies, to optimize the inter-provincial energy cooperation system, and to improve inter-provincial energy distribution, so as to achieve a "win–win" effect of economic growth and energy transition.

This article still has some limitations. Firstly, this article constructed a system of evaluation indicators for NDI by combining existing research and data availability. However, the current representations of NDI have not yet been unified, so the system of indicators for NDI still needs improvement. Secondly, this paper chose mediating mechanisms related to green development, but we did not explore other potential influencing mechanisms. Due to the availability of data, it is very hard to enumerate all potential mechanisms, which will be the focus of future research and needs to be further explored.

Author Contributions: Conceptualization, L.F.; methodology, L.F.; formal analysis, Q.M., L.F. and J.Z.; investigation, L.F.; data curation, Q.M., L.F. and J.Z.; writing—original draft preparation, L.F.; Y.Z. and M.J. writing—review and editing, Q.M., L.F. and J.Z.; project administration, L.F.; experiments, Y.Z. and M.J. All authors have read and agreed to the published version of the manuscript.

Funding: This research received no external funding.

Data Availability Statement: Not applicable.

Conflicts of Interest: The authors declare no conflict of interest.

References

1. Sharma, V.; Greig, C.; Lant, P. What is stopping India's rapid decarbonisation? Examining social factors, speed, and institutions in Odisha. *Energy Res. Soc. Sci.* **2021**, *78*, 102117. [CrossRef]
2. Wei, C. CO_2 marginal abatement cost and determinants-An empirical analysis of sample cities in China. *China Econ.* **2015**, *10*, 102. [CrossRef]
3. Ahmad, T.; Zhang, D. A critical review of comparative global historical energy consumption and future demand: The story told so far. *Energy Rep.* **2020**, *6*, 1973–1991. [CrossRef]
4. Henderson, J.; Mitrova, T. Implications of the global energy transition on Russia. In *The Geopolitics of the Global Energy Transition*; Springer: Cham, Swithland, 2020; pp. 93–114. [CrossRef]

5. Solomon, B.D.; Krishna, K. The coming sustainable energy transition: History, strategies, and outlook. *Energy Policy* **2011**, *39*, 7422–7431. [CrossRef]
6. Chandra, S.; Shirish, A.; Srivastava, S.C. Theorizing technological spatial intrusion for ICT enabled employee innovation: The mediating role of perceived usefulness. *Technol. Forecast. Soc. Chang.* **2020**, *161*, 120320. [CrossRef]
7. Yang, H.; Liu, F.; Zhang, L. GIP Evaluation and Path Improvement of Technological SMEs Based on Digital Information Technology. *Mob. Inf. Syst.* **2022**, *2022*, 1574267. [CrossRef]
8. Nguyen, T.T.; Pham, T.A.T.; Tram, H.T.X. Role of information and communication technologies and innovation in driving carbon emissions and economic growth in selected G-20 countries. *J. Environ. Manag.* **2020**, *261*, 110162. [CrossRef]
9. Alhassan, M.D.; Adam, I.O. The effects of digital inclusion and ICT access on the quality of life: A global perspective. *Technol. Soc.* **2021**, *64*, 101511. [CrossRef]
10. Sun, H.; Kim, G. The composite impact of ICT industry on lowering carbon intensity: From the perspective of regional heterogeneity. *Technol. Soc.* **2021**, *66*, 101661. [CrossRef]
11. Oughton, E.; Tyler, P.; Alderson, D. Who's Superconnected and Who's Not? Investment in the UK's Information and Communication Technologies (ICT) Infrastructure. *Infrastruct. Complex.* **2015**, *2*, 6. [CrossRef]
12. Galvin, R. The ICT/electronics question: Structural change and the rebound effect. *Ecol. Econ.* **2015**, *120*, 23–31. [CrossRef]
13. Zheng, J.; Wang, X. Can mobile information communication technologies (ICTs) promote the development of renewables?-evidence from seven countries. *Energy Policy* **2021**, *149*, 112041. [CrossRef]
14. Kallal, R.; Haddaji, A.; Ftiti, Z. ICT diffusion and economic growth: Evidence from the sectorial analysis of a periphery country. *Technol. Forecast. Soc. Chang.* **2021**, *162*, 120403. [CrossRef]
15. Chung, H. ICT investment-specific technological change and productivity growth in Korea: Comparison of 1996–2005 and 2006–2015. *Telecommun. Policy* **2018**, *42*, 78–90. [CrossRef]
16. Tang, C.F.; Tan, E.C. Exploring the nexus of electricity consumption, economic growth, energy prices and technology innovation in Malaysia. *Appl. Energy* **2013**, *104*, 297–305. [CrossRef]
17. Lange, S.; Pohl, J.; Santarius, T. Digitalization and energy consumption. Does ICT reduce energy demand? *Ecol. Econ.* **2020**, *176*, 106760. [CrossRef]
18. Shabani, Z.D.; Shahnazi, R. Energy consumption, carbon dioxide emissions, information and communications technology, and gross domestic product in Iranian economic sectors: A panel causality analysis. *Energy* **2019**, *169*, 1064–1078. [CrossRef]
19. Mourshed, M.; Robert, S.; Ranalli, A.; Messervey, T.; Reforgiato, D.; Contreau, R.; Becue, A.; Quinn, K.; Rezgui, Y.; Lennard, Z. Smart Grid Futures: Perspectives on the Integration of Energy and ICT Services. *Energy Procedia* **2015**, *75*, 1132–1137. [CrossRef]
20. Bris, M.; Pawlak, J.; Polak, J.W. How is ICT use linked to household transport expenditure? A cross-national macro analysis of the influence of home broadband access. *J. Transp. Geogr.* **2017**, *60*, 231–242. [CrossRef]
21. Hilty, L.M. *Information Technology and Sustainability*; Books on Demand: Nordstedt, Germany, 2008. [CrossRef]
22. Ishida, H. The effect of ICT development on economic growth and energy consumption in Japan. *Telemat. Inform.* **2015**, *32*, 79–88. [CrossRef]
23. International Renewable Energy Association. *Future of Wind: Deployment, Investment, Technology, Grid Integration and Socio-economic Aspects*; International Renewable Energy Agency: Abu Dhabi, United Arab Emirates, 2019.
24. O'Dwyer, E.; Pan, I.; Acha, S.; Gibbons, S.; Shah, N. Modelling and evaluation of multi-vector energy networks in smart cities. In Proceedings of the International Conference on Smart Infrastructure and Construction 2019 (ICSIC) Driving data-informed decision-making, Cambridge, UK, 8–10 July 2019; ICE Publishing: London, UK, 2019; pp. 161–168. [CrossRef]
25. Duch-Brown, N.; Rossetti, F. Digital platforms across the European regional energy markets. *Energy Policy* **2020**, *144*, 111612. [CrossRef]
26. Korkmaz, P.; Gardumi, F.; Avgerinopoulos, G.; Blesl, M.; Fahl, U. A comparison of three transformation pathways towards a sustainable European society—An integrated analysis from an energy system perspective. *Energy Strat. Rev.* **2020**, *28*, 100461. [CrossRef]
27. Thongmak, M. Youths' Green Information and Communications Technology Acceptance and Implications for the Innovation Decision Process. *Electron. Green J.* **2016**, *1*. [CrossRef]
28. Moyer, J.D.; Hughes, B.B. ICTs: Do they contribute to increased carbon emissions? *Technol. Forecast. Soc. Chang.* **2012**, *79*, 919–931. [CrossRef]
29. Yan, Z.; Shi, R.; Yang, Z. ICT Development and Sustainable Energy Consumption: A Perspective of Energy Productivity. *Sustainability* **2018**, *10*, 2568. [CrossRef]
30. Haftu, G.G. Information communications technology and economic growth in Sub-Saharan Africa: A panel data approach. *Telecommun. Policy* **2019**, *43*, 88–99. [CrossRef]
31. Johnstone, N.; Haščič, I.; Popp, D. Renewable energy policies and technological innovation: Evidence based on patent counts. *Environ. Resour. Econ.* **2010**, *45*, 133–155. [CrossRef]
32. Østergaard, P.A.; Duic, N.; Noorollahi, Y.; Kalogirou, S.A. Recent advances in renewable energy technology for the energy transition. *Renew. Energy* **2021**, *179*, 877–884. [CrossRef]
33. Motz, A. Consumer acceptance of the energy transition in Switzerland: The role of attitudes explained through a hybrid discrete choice model. *Energy Policy* **2021**, *151*, 112152. [CrossRef]

34. Yang, X. FinTech in Promoting the Development of Green Finance in China against the Background of Big Data and Artificial Intelligence. In Proceedings of the 2020 4th International Seminar on Education Innovation and Economic Management (SEIEM), Penang, Malaysia, 23–24 May 2020; Francis Academic Press: London, UK, 2020. [CrossRef]
35. Qing, H.G. Green information technology government regulation components: Improving Indonesia green information technology. *J. Theor. Appl. Inf. Technol.* **2019**, *97*, 4467–4477.
36. Zhao, Y.; Ke, J.; Ni, C.C.; McNeil, M.; Khanna, N.Z.; Zhou, N.; Fridley, D.; Li, Q. A comparative study of energy consumption and efficiency of Japanese and Chinese manufacturing industry. *Energy Policy* **2014**, *70*, 45–56. [CrossRef]
37. United Nations Environment Programme (UNEP). *Adaptation Gap Report 2020-Executive Summary*; United Nations Environment Programme: Nairobi, Kenya, 2021.
38. O'Mahony, M.; Vecchi, M. Quantifying the Impact of ICT Capital on Output Growth: A Heterogeneous Dynamic Panel Approach. *Economica* **2005**, *72*, 615–633. [CrossRef]
39. Li, W.; Fan, J.; Zhao, J. Has green finance facilitated China's low-carbon economic transition? *Environ. Sci. Pollut. Res.* **2022**, *29*, 57502–57515. [CrossRef] [PubMed]
40. Navarro, M.P.; Catalão-Lopes, M. A structured financial product applied to renewable energies. *Green Financ.* **2019**, *1*, 82–93. [CrossRef]
41. Huang, J.; Xia, J. Regional Competition, Heterogeneous Factors and Pollution Intensity in China: A Spatial Econometric Analysis. *Sustainability* **2016**, *8*, 171. [CrossRef]
42. Dong, F.; Yu, B.; Zhang, J. What Contributes to Regional Disparities of Energy Consumption in China? Evidence from Quantile Regression-Shapley Decomposition Approach. *Sustainability* **2018**, *10*, 1806. [CrossRef]
43. Hu, A.G. *Innovating Green Development in China*; Springer: Berlin/Heidelberg, Germany, 2012.
44. Soma, K.; Termeer, C.J.; Opdam, P. Informational governance—A systematic literature review of governance for sustainability in the Information Age. *Environ. Sci. Policy* **2016**, *56*, 89–99. [CrossRef]
45. Zhang, H. The green paradox puzzle: Interpretation from the perspective of local government competition. *J. Financ. Econ-Nomics* **2014**, *40*, 114–127.
46. Galloway, E.; Johnson, E.P. Teaching an old dog new tricks: Firm learning from environmental regulation. *Energy Econ.* **2016**, *59*, 1–10. [CrossRef]
47. Zheng, Y.; Peng, J.; Xiao, J.; Su, P.; Li, S. Industrial structure transformation and provincial heterogeneity characteristics evolution of air pollution: Evidence of a threshold effect from China. *Atmos. Pollut. Res.* **2020**, *11*, 598–609. [CrossRef]
48. Huang, S.-Z. The effect of natural resources and economic factors on energy transition: New evidence from China. *Resour. Policy* **2022**, *76*, 102620. [CrossRef]
49. Lu, H.; Peng, J.; Lu, X. Do Factor Market Distortions and Carbon Dioxide Emissions Distort Energy Industry Chain Technical Efficiency? A Heterogeneous Stochastic Frontier Analysis. *Energies* **2022**, *15*, 6154. [CrossRef]
50. Xiang, H.; Ch, P.; Nawaz, M.A.; Chupradit, S.; Fatima, A.; Sadiq, M. Integration and economic viability of fueling the future with green hydrogen: An integration of its determinants from renewable economics. *Int. J. Hydrog. Energy* **2021**, *46*, 38145–38162. [CrossRef]
51. Mohammed, I.; Lenshie, N.E. Political Economy of Resource Curse and Dialectics of Crude Oil Dependency in Nigeria. *Int. Bus. Manag.* **2017**, *14*, 33–44. [CrossRef]
52. Teklemariam, M.H.; Kwon, Y. Reducing internet demand-side gap improves digital inclusion in low-income coun-tries:-analysis that is more comprehensive. In Proceedings of the 22nd ITS Biennial Conference, Seoul 2018. Beyond the Boundaries: Challenges for Business, Policy and Society, Seoul, Korea, 24–27 June 2018; International Telecommunications Society (ITS): Calgary, AB, Canada, 2018.
53. Daly, H.E. Toward some operational principles of sustainable development. *Ecol. Econ.* **1990**, *2*, 97–102. [CrossRef]
54. Wu, Y.; Wang, J.; Ji, S.; Song, Z. Renewable energy investment risk assessment for nations along China's Belt & Road Initiative: An ANP-cloud model method. *Energy* **2020**, *190*, 116381. [CrossRef]
55. Naz, L.; Ali, A.; Fatima, A. International competitiveness and ex-ante treatment effects of CPEC on household welfare in Pakistan. *Int. J. Dev. Issues* **2018**, *17*, 168–186. [CrossRef]
56. Sun, X. Non-energy factors and their influences in energy cooperation between China and Middle East countries. *Arab. World Stud.* **2022**, *4*, 21–39.
57. Caldés, N.; del Río, P.; Lechón, Y.; Gerbeti, A. Renewable Energy Cooperation in Europe: What Next? Drivers and Barriers to the Use of Cooperation Mechanisms. *Energies* **2018**, *12*, 70. [CrossRef]
58. Adebayo, T.S.; Awosusi, A.A.; Bekun, F.V.; Altuntaş, M. Coal energy consumption beat renewable energy consumption in South Africa: Developing policy framework for sustainable development. *Renew. Energy* **2021**, *175*, 1012–1024. [CrossRef]
59. Zhao, X. Research on the effect of technological innovation of NDI. *Stat. Res.* **2022**, *39*, 80–92.
60. Hadfield, P.; Cook, N. Financing the Low-Carbon City: Can Local Government Leverage Public Finance to Facilitate Equitable Decarbonisation? *Urban Policy Res.* **2019**, *37*, 13–29. [CrossRef]
61. Saygin, D.; Kempener, R.; Wagner, N.; Ayuso, M.; Gielen, D. The Implications for Renewable Energy Innovation of Doubling the Share of Renewables in the Global Energy Mix between 2010 and 2030. *Energies* **2015**, *8*, 5828–5865. [CrossRef]
62. Yan, Y.; Liu, T.; Wang, N.; Yao, S. Urban sprawl and fiscal stress: Evidence from urbanizing China. *Cities* **2022**, *126*, 103699. [CrossRef]

63. Sinn, H.-W. Public policies against global warming: A supply side approach. *Int. Tax Public Financ.* **2008**, *15*, 360–394. [CrossRef]
64. Peng, B.; Chen, H.; Elahi, E.; Wei, G. Study on the spatial differentiation of environmental governance performance of Yangtze river urban agglomeration in Jiangsu province of China. *Land Use Policy* **2020**, *99*, 105063. [CrossRef]
65. Liu, J.; Xuan, K.; Xie, N.; Zhang, J.; Wang, X.; Yu, Z.; Wang, W. Effects of urbanisation on regional water consumption in China. *J. Hydrol.* **2022**, *609*, 127721. [CrossRef]
66. Taghizadeh-Hesary, F.; Rasoulinezhad, E. Analyzing Energy Transition Patterns in Asia: Evidence From Countries With Different Income Levels. *Front. Energy Res.* **2020**, *8*, 162. [CrossRef]
67. Acheampong, A.O.; Boateng, E.; Amponsah, M.; Dzator, J. Revisiting the economic growth–energy consumption nexus: Does globalization matter? *Energy Econ.* **2021**, *102*, 105472. [CrossRef]
68. Li, F.; Shang, Y.; Xue, Z. The impact of foreign direct investment on my country's green development: Based on data verification of 260 prefecture-level cities in China. *Econ. Issues* **2022**, *19*, 12183. [CrossRef]
69. Oughton, E. Policy options for digital infrastructure strategies: A simulation model for broadband universal service in Africa. *arXiv* **2021**, arXiv:2102.03561. [CrossRef]
70. Shen, F.; Liu, B.; Luo, F.; Wu, C.; Chen, H.; Wei, W. The effect of economic growth target constraints on green technology innovation. *J. Environ. Manag.* **2021**, *292*, 112765. [CrossRef] [PubMed]
71. Wang, H.; Jiang, L.; Duan, H.; Wang, Y.; Jiang, Y.; Lin, X. The Impact of Green Finance Development on China's Energy Structure Optimization. *Discret. Dyn. Nat. Soc.* **2021**, *2021*, 2633021. [CrossRef]
72. Xu, Q.; Zhong, M.; Li, X. How does digitalization affect energy? International evidence. *Energy Econ.* **2022**, *107*, 105879. [CrossRef]
73. Du, W.; Li, M. Assessing the impact of environmental regulation on pollution abatement and collaborative emissions reduction: Micro-evidence from Chinese industrial enterprises. *Environ. Impact Assess. Rev.* **2020**, *82*, 106382. [CrossRef]
74. Zhao, J. The Belt and Road Energy Cooperation. In *Reshaping the Economic Cooperation Pattern of the Belt and Road Initiative*; Palgrave Macmillan: Singapore, 2021; pp. 151–169.
75. Hu, Y.; Li, Y.; Sun, J.; Zhu, Y.; Chai, J.; Liu, B. Towards green economy: Environmental performance of belt and road initiative in China. *Environ. Sci. Pollut. Res.* **2022**. [CrossRef]
76. Xu, H.; Cui, Q.; Sofield, T.; Li, F.M.S. Attaining harmony: Understanding the relationship between ecotourism and protected areas in China. *J. Sustain. Tour.* **2014**, *22*, 1131–1150. [CrossRef]
77. Zhang, K.Q.; Chen, H.H.; Tang, L.Z.; Qiao, S. Green Finance, Innovation and the Energy-Environment-Climate Nexus. *Front. Environ. Sci.* **2022**, *10*, 602. [CrossRef]

energies

MDPI

Article

Optimization of Exploration and Production Sharing Agreements Using the Maxi-Min and Nash Solutions

Saad Balhasan [1], Mohammed Alnahhal [2,*], Shahrul Shawan [1], Bashir Salah [3], Waqas Saleem [4] and Mosab I. Tabash [5]

1 Chemical and Petroleum Engineering Department, American University of Ras Al Khaimah, Ras Al Khaimah P.O. Box 10021, United Arab Emirates
2 Mechanical and Industrial Engineering Department, American University of Ras Al Khaimah, Ras Al Khaimah P.O. Box 10021, United Arab Emirates
3 Department of Industrial Engineering, College of Engineering, King Saud University, P.O. Box 800, Riyadh 11421, Saudi Arabia
4 Department of Mechanical and Manufacturing Engineering, Institute of Technology, F91 YW50 Sligo, Ireland
5 College of Business, Al Ain University, Al Ain P.O. Box 64141, United Arab Emirates
* Correspondence: mohammed.alnahhal@aurak.ac.ae; Tel.: +971-7-246-8748

Abstract: Cooperation between supply chain partners in the oil industry is essential, especially when oil prices suffer from fluctuations that affect the profitability of each party. An essential task in oil field development projects is to create an optimum agreement between the national oil company and the international oil company to guarantee agreement optimization. In this paper, the national oil company is the first party (FP) and the international oil company is the second party (SP). The paper's purpose is to investigate the use of game theory to obtain the best agreement between the FP and SP in order to enhance the cooperation and reduce conflict. In this paper, Nash and Maxi-min solutions have been applied for the first time in a special type of petroleum agreement, called exploration and production sharing agreements (EPSA). This is conducted for a case study in Libya. The study considers nine negotiation factors (issues) in the EPSA, which are the share percent, the four "A" factors, and the four "B" factors, which are usually affected by the fluctuations of oil prices; and the study investigates their effect on the total payoff function, the net present value (NPV), and internal rate of return (IRR) for both parties. The Maxi-min solution has shown an improvement in the NPV and IRR of the SP, where NPV increased from USD 148 million to USD 195 million, and IRR from 15.65% to 17.01%. The Nash solution has shown a little more improvement than the Maxi-min solution in the NPV and IRR for the SP, where the NPV and IRR have increased from USD 148 million to USD 222 million and from 15.65% to 17.94%, respectively.

Keywords: oil fields; oil companies; negotiation; game theory; Maxi-min solution; Nash solution; agreement optimization

Citation: Balhasan, S.; Alnahhal, M.; Shawan, S.; Salah, B.; Saleem, W.; Tabash, M.I. Optimization of Exploration and Production Sharing Agreements Using the Maxi-Min and Nash Solutions. *Energies* **2022**, *15*, 8970. https://doi.org/10.3390/en15238970

Academic Editors: Jiachao Peng, Le Wen, Jianzhong Xiao, Ming Yi and Mingyue (Selena) Sheng

Received: 2 November 2022
Accepted: 25 November 2022
Published: 27 November 2022

1. Introduction

Negotiation is described as a process in which two or more parties negotiate or cooperate in order to reach an agreement. Systematic studies of the primary sources of negotiation literature have been published by Kemper and Kemper [1]. The origins of negotiation research can be traced back to game theory. Raiffa's dissertation, which is included in his book *"The Art and Science of Negotiation"*, focuses on game theory to explore negotiation theories' strategic choices. He claims that the effectiveness of negotiations is contingent on specific decisions [2]. Pruitt and Carnevale [3] addressed the social conflict negotiation outlines of the dominant normative negotiating paradigm's faulty principles; these traditional models assume that there are only two negotiating sides, each structured to maximize self-interest.

The use of specific methodology and scientific research to identify the best alternatives is an important aspect of the process of evaluating viable investment opportunities and assisting decision making. The general characteristics of petroleum project evaluation are comparable to those of other industries. There are, however, some unique and distinct challenges, most of which are related to the nature and conditions of petroleum projects and necessitate specialized knowledge and experience. [4].

In comparison to other oil agreements, exploration and production sharing agreements (EPSA) are currently more widespread. Many EPSA conditions allowed for negotiations between the national oil company, which is the first party (FP), and the international oil company, which is second party (SP). The FP retains rights to petroleum resources and production under the EPSA, but the SP receives a part of hydrocarbon production in exchange for services rendered [5]. The EPSA is used to split the profits from developed oil and gas fields' hydrocarbon output. The EPSA allows for a variety of profit-sharing arrangements between the two parties. Production share, profit split, production rate, bonus, discounted cash flow, royalty, and income tax are some of the most used methods [6]. In profit oil split, most EPSAs use a production-based sliding scale and R-factor method. Around 75% of EPSAs in the world use a sliding scale based on daily production and annual SP investment [6].

In this research, the Libyan EPSA IV model was applied to an oil field with secondary recovery and water injection. The development costs are split evenly between the FP and the SP (50% −50%). The running costs are split among the partners based on their output shares (production share). All costs for exploration, appraisal, and development can be recovered from the SP's production share. Taxes, royalties, and other fees are not applied to the SP [7,8]. To estimate the net present value (NPV) and internal rate of return (IRR) of this field, an economic model was created in Excel. Compared to the FP, the SP's NPV and IRR of the economic model are both low. The FP's main goal is to maximize profit from existing oil and gas reserves. The SP intends to increase oil output while lowering expenses and increasing profit. The SP has highlighted concerns about potential conflicts of interest between the FP and the SP. This issue arose as a result of the SP's unsatisfactory return of the agreement's earnings.

A smooth process of agreement between the two parties might face some challenges in determining the terms in the contract. This is especially important in the time of heavy fluctuations of prices of oil. Every party wants the best agreement terms to maximize its profit. Therefore, a fair agreement based on a certain methodology is necessary. The methodology needs to be practical and easy to understand by practitioners in the two parties. This study concentrates on the variables that are reflective to the changes in the prices of oil. Based on Nash and Maxi-min solutions, the study proposes a method that can be applied with Excel spreadsheets to make it easy for the two parties to accomplish. To validate the methodology and give full details of its steps, a real case study was presented to show the effect of the proposed methodology. The effect of the production share, A factors, and B factors on the economic indicators NPV and IRR for the two parties (government and international company) was identified in the literature. The agreement on the levels of these factors needs to be investigated. In this study, the key nine negotiable factors are used and thoroughly examined using two game theory models in order to assist decision makers in determining the best course of action for each of them. Furthermore, a strategy for resolving the conflict between FP and SP has been developed to help remove conflict as a barrier to the development of oil fields.

Two solutions approaches have been used in this paper to resolve the conflict and optimize the agreement output. The Nash bargaining solution and Maxi-min solution are used to enhance the payoff and eventually the SP's NPV and IRR. The concentration on the SP is because its margin for profit is much less than that for the FP. However, there is a threshold for the minimum score that the FP can demand in negotiation, and this threshold was taken into consideration. The two methods show a significant improvement of the NPV and IRR. Agreement optimization for the parties has been achieved. The study proposes a practical methodology that is deep and yet easy to be applied by practitioners using Excel Solver. The study starts from the situation that currently exists and proposes a different way for more gain that guarantees the basic requirements of the two parties.

For the FP and SP alike, the division of economic rent is the primary focus of concern. It is the driving force behind a fiscal system and the focal point of negotiations and, at times, tensions and controversies [9].

The SP wants to maximize the value of their assets. They examine investment possibilities around the world and assess their relative risk–reward profiles using economic indicators. Oil companies monitor the revenue generated by assets throughout their economic lives to ensure that the capital investment and expenditures have been covered and the return on capital is compatible with the risk associated with the asset and the corporation's strategic objectives. The host government is interested in determining if a fiscal system achieves its goals. To do so, host governments utilize economic and system measures at the project level to analyze whether the project's financial and social benefits are commensurate with the project's risk level and the government's sector policy objectives. At the country level, host governments assess the influence of the oil sector's overall revenue flow on important macroeconomic variables (mainly inflation, GDP growth, the balance of payments) [10].

Better negotiation results will lead eventually to better NPV and IRR. The NPV is the difference between the present value of the investment's cash outflow and the present value of the project's cash inflow. Technically, when the cash flow of an activity is discounted at a certain given discount rate, either a positive or negative value is obtained, depending on the conditions. To obtain the maximum possible profit or benefit, the company will choose the activity with the highest NPV. The IRR is defined as the discount rate at which the sum of all future discounted cash flow present values equals zero. In the case of overseas investment, it will be significantly fair if an IRR of 13–17% is guaranteed to the SP. IRR becomes a more important profit indicator when its value is less than 20%. The IRR is thought to be considerably more significant for the SP than it is for the FP in oil development projects. This is the only issue that needs to be taken into account in EOR and offshore development projects where the IRR is crucial to the FP and SP. Therefore, the minimal IRR of 18% to 20% is taken into consideration as an economic criterion in the current study for economically recoverable resources [11].

The main contribution in this study is to include, for the first time, agreement variables (share, "A" factor, and "B" factor) in the decision-making process in EPSA agreements, using the two methods of Nash and Maxi-min solutions. This is done by investigating the effect of these variables on the profitability of the SP and FP. Moreover, a comparison is made between the proposed methodologies with the original strategy followed currently in the case study in Libya. In order to optimize their economic benefits, the two parties will decide which agreement factors to concentrate on during the agreement negotiation with the help of the proposed tool. Additionally, the NPV and IRR of the SP was significantly improved using the two new approaches without violating the profit requirements of the FP. Furthermore, the two fair solution approaches, Nash and Maxi-min, used in this research, for the first time in oil agreements, will offer a novel technique for further studies to modernize the current approaches being used in the negotiation of the equity split in the oil and gas industry to achieve agreement optimization.

The rest of this paper is organized as follows: After this introduction, the next section is about the literature review, which explains the previous related studies and the contribution of the current research. Then, the case study with full details is presented. Then, the methodology section explains the Nash and Maxi-min methods. Then, the section of results and analysis presents the results and main insights of the methods. Finally, the conclusion summarizes the main findings and recommendations for future research.

2. Literature Review

Applications of game theory in the oil and gas industry typically fall into one of three categories. The first category is competitive bidding, in which companies compete for a limited number of opportunities. The second type of partnership is a joint venture, in which a group of companies work together to implement a project or other opportunity. The third one is the negotiation that involves partners, clients, vendors, and governments, in which each side aims to secure the maximum possible share [12]. Game theory is known in the literature to be applied in the field of oil production and price [13]. However, none of the previous studies investigated game theory methods such as the Nash solution in EPSA agreements. For example, strategic exploration and production were derived jointly in a three-period subgame perfect equilibrium in a work by [14]. They found the subgame perfect Nash equilibrium in a game where firms compete not only in the output market but also in the exploration process. A game theoretic framework has been applied in a study by Willigers et al. [15] in the oil and gas industry, where the Nash equilibrium was used in the analysis. Esmaeili et al. [16] used a game theory approach to investigate the policies for Iran's oil and gas shared resources conflicts with Iraq and Qatar. The outcomes of mathematical models demonstrate how countries could devise an acceptable plan for utilizing their common resources. Langer et al. [17] used a partial-equilibrium global energy market model. The problem was modeled as a Generalized Nash Equilibrium (GNE) between non-cooperative players. They discovered that eliminating the US crude ban will benefit domestic producers by allowing them to sell their petroleum at global market prices rather than prices skewed by local constraints.

Tominac and Mahalec [18] created a game-theoretic framework for strategic production planning in petroleum refineries. The problem is expressed as non-cooperative potential games with Nash equilibria as solutions. According to game theory, the production planning choices are sound, and they can withstand changes in competition strategy. Moradinasab et al. [19] investigated the petroleum supply chain in light of sustainability and pricing challenges, and a model for a sustainable competitive petroleum supply chain was developed to reduce pollution while increasing profitability and job creation. Araujo and Leoneti [20] analyze relevant realistic and real-world oil and gas sector examples in the form of 2 × 2 strategic games, with the goal of investigating game theory methodologies to aid in the discussion and resolution of the major challenges encountered. They investigated the use of the Nash equilibrium and Max-min methods, plus other methods, to obtain solutions in different case studies.

Nicoletti and You [21] modeled the crude oil supply chain from oil well to refinery as a mixed-integer program that allows for competing objectives and interactions among various stakeholders. They applied the Stackelberg game theory. The crude oil refiner takes the lead and selects how much oil to buy in order to maximize profits while limiting the environmental impact of its operations. The profitability of investment in refinery development was investigated in a work by Babaei et al. [22], and the effects of the model on each agent were considered using a multi-agent method. Using a game theory approach, they discovered substantial investment problems with consequences for the future of the gasoline sector. Xue et al. [23] determined the optimum negotiation technique for oil corporations taking part in global oil and gas development projects. They created a model of bilateral bargaining and examined the variables that affect the equilibrium income ratios. Bidding order and information asymmetry are shown to be the two key influencing factors. The findings indicated that information asymmetry has no impact on the two parties' relative real income levels. Araujo and Leoneti [24] suggested using game theory to simulate and assess the stability of Brazil's regulatory framework for exploration and production. They suggested a method for modeling a multi-criteria group problem as a multi-criteria game and solved it by applying the Graph Model for Conflict Resolution methodology, to comprehend and measure the preferences of the players and find fair and stable solutions. Csercsik [25] constructed a simple game-theoretic model to capture the fundamental elements of the gas supply dilemma. The model was used to build a method for supply–security cooperation.

For more about game theory with application to oil production and price, the reader can refer to Ibrahim et al. [13]. The above studies investigated the use of game theory in the petroleum field in general. However, little was published on using game theory in petroleum contracts between the national company and the contractor. An example for that is the study by Keshavarz et al. [26], who investigated the Iranian petroleum contract fiscal regime using bargaining game theory for the purpose of guiding contract negotiators. The methodology presented depends on a certain type of contract (risk service contract) devolved by the Iranian government. Besides its narrow application field, the model presented is complex. Another study that investigated the game theory in petroleum contracts was the one by Dirani and Ponomarenko [27], who analyzed the production sharing contract system. The principle of win-win game theory was presented when the interests of the international oil company and the state are coordinated. However, they depended on a literature review and did not investigate the principle with data in detail. Moreover, none of the mentioned previous studies investigated the EPSA agreement. Therefore, the novelty of the current study is to investigate petroleum contracts using two types of game theory models, namely, the Nash and Maxi-min solutions, and propose two general models that can be applied easily in EPSA agreements and can be easy to understand. This is done with a real case study. To the best of the authors' knowledge, this study is the first one that investigates the application of a game theory method in EPSA agreements and their negotiation factors. The study depends on practical models that can be applied by companies to reduce the conflict between the two parties. Excel Solver was used because of its availability in every computer.

3. Case Study

The focus of the case study is to resolve the conflict between the Libyan National Oil Corporation (NOC) and an International Oil Company to develop the AA oil field. The National Oil Corporation was established in 1970. Its purpose is to organize petroleum development plans and to oversee the administration and financial operations of oil and gas enterprises. The NOC is in charge of all oil and gas exploration, production, and marketing both domestically and abroad through its subsidiaries (National Companies) or through agreements with foreign companies [28]. The NOC has plans to raise Libyan oil production capacity to 2 million barrels per day. The NOC highlighted its plans for exploration and production by the following steps:

1. Maximize the profit from each oil and gas agreement.
2. Minimize the SP share in any oil and gas agreement to obtain the highest revenue.

The SP has to bear a high portion of the risk. The SP expects to meet the benchmark economic criterion. In the proposed development scenario, the IRR did not reach the minimum limit.

To maintain the production plateau and boost the oil recovery factor, the AA oil field is expected to be developed by a water injection project. The two parties intend to drill 50 producing wells with a daily flow rate of roughly 60 thousand barrels to develop the field. This rate is likely to push the plateau out for another six years. The remaining four peripheral water injections will be drilled to guarantee the requisite oil rate and pressure are met. Table 1 shows the capital CAPEX and OPEX of the AA oil field. The total production is expected to reach 219 million barrels by 2037, according to projections. A three-phase separator is recommended in the field due to the relatively high gas–oil ratio of 800 SCF/STB and water output. The condensate output of the field is estimated to be 30 STB/MSCF. As a result, a gas plant would be required to remove liquid hydrocarbon by-product (LHP) from the field.

Table 1. The capital expenditure and operating costs of AA field.

Cost Type	Value, USD MM
CAPEX	569.18
OPEX	464.5
Other Costs	91.35
Total Costs	1125.03

A coded spreadsheet model was utilized to estimate the profitability indicators of NPV and IRR for this scenario based on the AA field data. The purpose of this coded model is to create a decision-making model for the initial development scenario.

This field development scenario assumes that the field was created using primary and secondary recovery methods, as well as water injection. Oil field size, oil prices, gas prices, LHP prices, and others are decision factors in this regard. Table 2 shows the assumptions for the decision factors of the AA field in the economic model.

Table 2. The capital expenditure and operating costs of AA field.

Project Variable	Value
Original Oil in Place	1 billion
The initial production	60,000 STB/D
Plateau time	6 years
The decline model	Hyperbolic
The annual decline rate	25%
The hyperbolic constant	0.6
The oil price (escalated)	USD 65/barrel
The HLP price (escalated)	USD 75/barrel
The gas price (escalated0	USD 5/MMBTU
The discount rate	10%
The inflation rate	2%
The borrowed money (50% of the CAPEX)	USD 321 million
The payment period	5 years
The loan interest rate	7%
The SP Production Share	15%

The Equity Split Mechanism

In 2004, the Libyan EPSA IV modified version was launched. The agreement requires the SP to assume full responsibility for all exploration costs. The FP pays the entire share of operational costs (equivalent to its contractual share, 85% to 90%) but only half of development costs. Once production begins, the SP sets their proportion of share at 10% to 15% of total production in order to recover their share of the exploration and development costs. The term "production share" refers to this percentage. Furthermore, according to the "A" and "B" factors, the excess profit oil (the remaining oil from the second party's share of production "10% to 15%") is shared between the two parties. As will be explained later, the "A" and "B" factor values are a matter of negotiation between the two parties. Signature and Production Bonuses must be paid by the second party. The first party, on the other hand, pays the income tax of the second party from its share of the revenue to the Libyan government. Furthermore, the second party is exempted from customs duties under Libyan petroleum law. The following are specific rules in EPSA IV:

1. Similar to EPSA III, except with added Gas and LHP Clauses.
2. SP is entirely responsible for exploration expenditures.
3. CAPEX is split 50/50 between the two parties.
4. SP's percent of output provided for SP cost recovery.
5. OPEX is shared according to the production share.
6. There is no royalty and no tax paid by a third party.
7. The original "B" factors are as shown in Table 3, and they are a step function of field oil output. The results of this study provide better settings as will be shown later.
8. Just like the "B" factors, the original "A" factors are obtained to compare the results of this study to them. They are shown in Table 4, and they are a step function of the R ratio. The two parties' negotiating parameters include both "A" and "B" factors.

Table 3. Initial settings of Production Rate and Production Index, B factor.

Production Rate (bbl/Day)	Production Index B
1–20,000	0.95
20,001–30,000	0.8
30,001–60,000	0.6
60,001–85,000	0.45
>85,000	0.2

Table 4. Initial settings of A Factor and R Ratio.

R Ratio	A Factor
1.0–1.5	0.9
1.5–3.0	0.8
3.0–4.0	0.6
>4.0	0.4

The net cash flow (NCF) in the EPSA IV model can be found by using the following equations [28,29].

$$\begin{aligned} \text{FP NCF} = {} & [(\text{FP Share\%} * \text{oil production*price}) + (\text{excess profit, oil}) - (\text{SP excess profit, oil})] \\ & + [(\text{FP Share\%} * \text{LHP production*price}) + (\text{excess profit, LHP}) - (\text{SP excess profit, LHP})] \\ & + [(\text{FP Share\%} * \text{gas production*price}) + (\text{excess profit, gas}) \\ & - (\text{SP excess profit, gas})] + \text{Production Bonus} - \text{CAPEX} - \text{OPEX} \end{aligned} \tag{1}$$

$$\begin{aligned} \text{SP NCF} = {} & [(\text{SP Share\%} * \text{oil production*price}) - (\text{excess profit, oil}) \\ & + (\text{SP excess profit, oil})] + [(\text{SP Share\%} * \text{LHP production*price}) \\ & - (\text{excess profit, LHP}) + (\text{SP excess profit, LHP})] \\ & + [(\text{SP Share\%} * \text{gas production*price}) - (\text{excess profit, gas}) \\ & + (\text{SP excess profit, gas})] - \text{CAPEX} - \text{OPEX} - \text{Production Bonus} - \text{Capital cost} \end{aligned} \tag{2}$$

The net present value represents the discounted values of future cash inflows and outflows related to a specific project. The project lifetime is 29 years. After finding the NCF based on the above equation, NCF of the SP was deflated. Then, IRR was estimated using the function of IRR in Excel. Then, the NPV was determined by taking the sum of the negative and positive cash flows and discounting the deflated NCF (from the IRR) by using the NPV function.

4. Methodology

The Nash bargaining solution is an optimization procedure used to maximize the product of the payoffs. Almost all bargaining, according to Nash, is a method of achieving and distributing benefits. A collection of possible variations of the division of the jointly obtained benefits from all possible arrangements of the subjects can be considered as such a negotiation scenario, with the point of conflict "d" determining the subset of the set "S" within which the solution will be sought, see Figure 1.

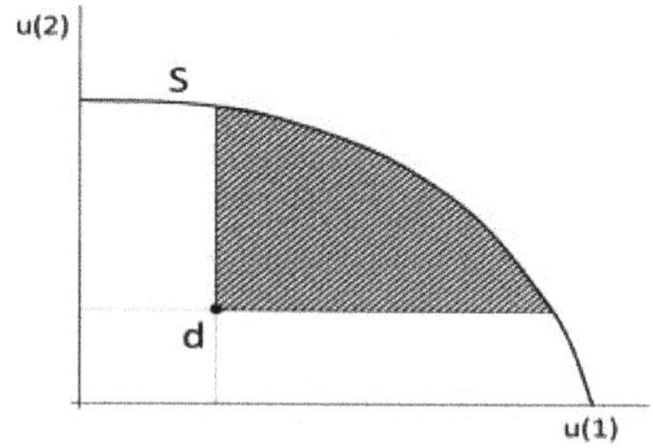

Figure 1. Negotiation as a Nash's bargaining problem.

The point of contention defines a compromise to which all negotiating sides agree without further discussion. Alternatively, the negotiation is the product of both parties' alternatives—thus, a compromise for a lower profit than that specified by point is not worth it. The point of contention may also be found at the crossroads of axes (x, y) if neither party can come to an agreement and there are no alternative options, as even minor improvements are beneficial for both parties [30].

Nash's bargaining solution is a precise solution based on a number of assumptions. Perfect details, fair negotiating skills, knowledge of the power of negotiation, and so on are examples of these. Nash suggests a solution which is the so-called Nash product, which can be found using the formula:

$$Max\ [u_1\ (x^*) - u_1\ (x^0)][u_2\ (x^*) - u_2\ (x^0)] \quad (3)$$

where u_1 and u_2 correspond to utilities of the first and second subject, point x^0 is the benefit at the point of disagreement, and the point x^* relates to the point of interest. Thus, as a result, the formula shows the maximum benefit that entities can receive [30].

On the other hand, Maxi-min is used to maximize the minimum, to change the objective function of the agreement output from maximizing the product of the payoffs to FP and SP to maximizing the lesser of the two payoffs.

It is an optimization procedure used to maximize the minimum of the proportional of the potential (POP) of the FP and SP. The POP is the ratio of the "excess" to the difference between the maximum feasible value and the reservation value (RV). Excess is the difference between the obtained value by the method used and the RV. Later, these relationships are expressed using mathematical equations. The Excel 2016 Solver will be used to find the agreement that would maximize the FP's and SP's scores. Okoro et al. [31] used Excel to make their analysis of game theory, where they used the Maxi-min solution.

There are some similarities between the two methods (Nash and Maxi-min solutions). Therefore, some previous studies such as Araujo and Leoneti [20] and Turbay and Reyes [32] compared and investigated both of them. However, the objective function of both of them is different. The Nash equilibrium seeks the best possible strategic option when compared to the options of other players, and this is true for all players. The Maxi-min strategy, on the other hand, seeks payoffs that are at least as good as the worst payoff from any other strategy [33]. Robinson and Goforth [33] proposed a 2 × 2 strategic game classification based on the players' payoff-space representation, particularly in the understanding and interpretation of the Nash equilibrium and Maxi-min strategies. In the next section, the difference between the results of both of them is presented.

The Optimization Method Using Maxi-Min and Nash Solution Models

The methodology in this study depends on subjective estimation of the importance of the negotiable variables, and the score of each level of the variables for both parties. The two parties might set together to estimate the importance and the score. The following steps are used for the optimization method [2]:

1- Identifying the variables (issues) to be negotiated by the FP and SP. In the terminology of negotiations, issues are used to represent the negotiator factors that need to be set by both parties. In this study, issues and variables are used interchangeably to mean the same thing.

The two parties have to determine whether the share, "A" factors, and "B" factors need to be negotiated or just "A" and "B" factors.

2- Determining the best values of each variable

The FP and SP should list for each issue a set of best and possible resolutions. In this paper, larger variable values are usually for the advantage of the SP.

3- Determining the preferences and value tradeoffs

The FP and SP should ordinally rank their preferences for a different resolution level for each issue. Moreover, the two parties have to place the issues in rank order from the highest importance to the lowest importance. It is known in the literature and based on some equations that some variables have a larger effect on the NPV and IRR. The exact effect, however, depends on uncontrollable factors that are not easy to forecast such as future prices and inflation. Therefore, subjective numbers are used in this study based on the experience of the authors.

4- The additive scoring system

The FP and SP should score their issues preferences. It would rather go from the most important to least important. Additionally, it would rather go from the worst to the best choice.

5- Determining the reservation values (RV)

The FP and SP should decide what the lowest acceptable score (RV) for each bargaining issue is.

6- Finding efficient contracts

The Maxi-min and Nash solutions have been used in this paper to find the most efficient contracts. Firstly, the FP and SP are jointly going to negotiate contracts and select one contract for the nine variables (share, four "A" factors, and four "B" factors) using Full, Open, Truthful Exchange. Secondly, Excel Solver is used to find the results. Solver will try to find the best contract that would maximize the minimum of the FP and SP POP. Finally, Solver is used to find a fair contract based on the Nash solution by maximizing the product of excesses [2].

The following sets are needed:

I is the number of variables, in the case study it is 9

J_i is the number of options for the variable i

The following parameters are given:

y_{ij1} is the payoff (score) for the FP if option j is selected for the variable i

y_{ij2} is the payoff (score) for the FP if option j is selected for the variable i

The following variables are needed:

$$x_{ij} = \begin{cases} 1\ if\ option\ j\ is\ chosen\ for\ the\ variable\ i\ in\ the\ optimal\ solution \\ 0\ otherwise \end{cases}$$

The objective is to maximize the objective function

$$max\ Z = \left(\sum_{i=1}^{I} \sum_{j=1}^{J_i} y_{ij1} x_{ij} - vr_1 \right) \left(\sum_{i=1}^{I} \sum_{j=1}^{J_i} y_{ij2} x_{ij} - vr_2 \right) \tag{4}$$

Subject to:

$$S_{i1} = \max_{1 \le j \le J_i} y_{ij1} \qquad \forall\, i = 1..I \tag{5}$$

$$S_{i2} = \max_{1 \le j \le J_i} y_{ij2} \qquad \forall\, i = 1..I \tag{6}$$

$$\sum_{i=1}^{I} S_{i1} = 100 \tag{7}$$

$$\sum_{i=1}^{I} S_{i1} = 100 \tag{8}$$

$$x_{ij}\ binary$$

The objective function defined in Equation (4) is the product of the excesses for both parties, which are the surpluses for both of them. Equations (5) and (6) are to define the score or importance of each variable, which is the maximum possible payoff that the party can obtain if the best option can be obtained. Equations (7) and (8) are to force the summation of the payoffs for all the variables for each party to be 100. For Maxi-min, the equations will be different. The constraints from (5) to (8) are used in the second model. However, the objective function is changed. To further explain that, some variables are defined as follows:

M_{F1} and M_{F2} maximum feasible value for the first and second party, respectively.
E_1 and E_2 excess are the excess for the first and second party, respectively.
P_1 and P_2 are potential for the first and second party, respectively.

The new equations will be:

$$max\ Z_2 = min(POP_1, POP_2) \tag{9}$$

S.T.

$$E_1 = \left(\sum_{i=1}^{I} \sum_{j=1}^{J_i} y_{ij1} x_{ij} - vr_1 \right) \tag{10}$$

$$E_2 = \left(\sum_{i=1}^{I} \sum_{j=1}^{J_i} y_{ij2} x_{ij} - vr_2 \right) \tag{11}$$

$$P_1 = M_{F1} - vr_1 \tag{12}$$

$$P_2 = M_{F2} - vr_2 \tag{13}$$

$$POP_1 = \frac{E_1}{P_1} \tag{14}$$

$$POP_2 = \frac{E_2}{P_2} \tag{15}$$

The objective is to minimize POP for the two parties. As explained before, the POP value is the excess divided by potential, and both of them are defined in Equations (10)–(13). The second model is linear, and that means it is easier to solve.

5. Results and Analysis

In this section, the results obtained using the two used methods are compared with the initial results set by the two parties without using our methods. In the original setting and on the basis of EPSA IV, the SP's NPV for estimated reserves of 219 million barrels was estimated to be USD 148 million and the IRR was 15.65%. The FP's NPV was estimated to be USD 5386 million and the IRR was 409%. Later in this section, the comparison is made. The FP and SP have determined nine issues and options (share, four A factors, and four B factors) to be negotiated. The negotiation issues and options have been prepared by the FP and SP for the negotiation to improve the SP's economic indicators, see Table 5. The table contains four to five options for each one of the nine decision variables. The methodology presented in the paper tries to select the best options for each variable. The importance of the share is much larger than the other variables. The ranges shown in Table 5 are determined based on the experience of the decision makers in both parties.

Table 5. The nine issues and options for the NOC and IOC.

Negotiation Issues and Options	Possible Options	Values
Production Share	Option 1	10%
	Option 2	12%
	Option 3	15%
	Option 4	18%
	Option 5	20%
A Factor 1 When R = (1.0–1.5)	Option 1	0.90
	Option 2	0.92
	Option 3	0.94
	Option 4	0.96
	Option 5	0.98
A Factor 2 When R = (1.5–3.0)	Option 1	0.78
	Option 2	0.80
	Option 3	0.82
	Option 4	0.84
	Option 5	0.86
A Factor 3 When R = (3.0–4.0)	Option 1	0.55
	Option 2	0.60
	Option 3	0.65
	Option 4	0.70
	Option 5	0.75
A Factor 4 When R = (>4.0)	Option 1	0.35
	Option 2	0.40
	Option 3	0.45
	Option 4	0.50
	Option 5	0.53
B Factor 1 When Production (bbl/day) (1–20,000)	Option 1	0.85
	Option 2	0.90
	Option 3	0.95
	Option 4	0.98
B Factor 2 When Production (bbl/day) (20,001–30,000)	Option 1	0.70
	Option 2	0.75
	Option 3	0.80
	Option 4	0.85
B Factor 3 When Production (bbl/day) (30,001–60,000)	Option 1	0.55
	Option 2	0.60
	Option 3	0.65
	Option 4	0.70
B Factor 4 When Production (bbl/day) (60,001–85,000)	Option 1	0.40
	Option 2	0.45
	Option 3	0.50
	Option 4	0.53

5.1. Effect of the Production Share, A factors, and B factors of EPSA IV on the SP's NPV and IRR

Minimizing the production share, A factors, and B factors in the EPSA IV adversely affect the NPV and IRR of the SP. This effect may appear clearly in oil projects that require large capital for such development facilities by using secondary and tertiary recovery. The EPSA IV determines the production share, which is supposed to recover expenses of the SP and give it a reasonable percentage of profits. By limiting the production share to a small value, the risk to the SP to recover their capital is increased when the payback period is increased. So, "A" factors reduce the profit of the foreign investor in case of stopping the investment or investing in limited range.

The positive impact of the value of investment on the profit of the SP only appears in the period of investment, which is the first period of the project. So, the SP will obtain the highest return from the profit oil when the "A" factors are kept at higher values. "B" factors are directly affected by the production rate, where a higher production rate will decrease the value of "B" factors and thus decrease the value of the oil profit and ultimately negatively impact the SP's produced share. The decline in the value of B factors due to increasing production gives a negative indicator to the SP and makes it not motivated to increase the production rate. The SP must negotiate the "B" factors that are dominated by the plateau of the production profile. On the other hand, the FP wants to minimize the benefit of the SP by minimizing the production share, "A" factors, and "B" factors [34]. The generated options, score, negotiation score for each issue, and the reservation value of the interest deal of the FP and the SP are shown in Table 6.

Table 6. Ranking issues by importance by the FP and SP.

Pr. Ranking	Issue	Pos. Resolution	FP		SP	
			Determined Value	Score	Determined Value	Score
1	Production Share, %	Option 1	60	60	20	
		Option 2	50		30	
		Option 3	40		40	
		Option 4	30		55	
		Option 5	20		60	60
2	B Factor 3	Option 1	10	10	6	
		Option 2	8		8	
		Option 3	6		10	
		Option 4	4		12	12
3	B Factor 1	Option 1	9	9	4	
		Option 2	8		6	
		Option 3	6		8	
		Option 4	4		10	10
4	A Factor 1	Option 1	6	6	2	
		Option 2	5		3	
		Option 3	4		4	
		Option 4	3		5	
		Option 5	2		6	6
5	A Factor 2	Option 1	5	5	1	
		Option 2	4		2	
		Option 3	3		3	
		Option 4	2		4	
		Option 5	1		5	5
6	A Factor 3	Option 1	5	5	1	
		Option 2	4		2	
		Option 3	3		3	
		Option 4	2		4	
		Option 5	1		5	5
7	B Factor 2	Option 1	3	3	0	
		Option 2	2.5		0	
		Option 3	2		1	
		Option 4	1		2	2
8	A Factor 4	Option 1	1	1	0	
		Option 2	0.5		0	
		Option 3	0		0	
		Option 4	0		0	
		Option 5	0		0	0

Table 6. *Cont.*

Pr. Ranking	Issue	Pos. Resolution	FP Determined Value	FP Score	SP Determined Value	SP Score
9	B Factor 4	Option 1	1	1	0	
		Option 2	0.5		0	
		Option 3	0		0	
		Option 4	0		0	0
			sum	100	sum	100

5.2. Nash Solution

The RV value for both parties must be determined at first. In this study, we assume it is 35 for the FP and 65 for the SP. The negotiation score output of FP and SP from the Nash solution is summarized in Table 7. This result was obtained with the assistance of Solver. Excel Solver is a unique mathematical program that operates within Excel. In Figure 2, the Solver dialogue box maximizes the objective of the product of the FP and SP excess. The product of excess is increased from 296, with the original settings, to 506. The formulation and solution of the problem that maximizes the sum of product of the FP and SP are given in Tables 7 and 8. The negotiation based on the Nash solution yields a solution for the FP and SP with the following production share, A factor, and B factor: production share at 15%, B3 at 0.65, B1 at 0.98, A1 at 0.98, A2 at 0.84, A3 at 0.75, B2 at 0.85, A4 at 0.35, and B4 at 0.40, see Table 8.

Table 7. Nash solution of the negotiation score of FP and SP from the Solver software.

Issue	Possible Options	Optimal Option	FP D. Value	FP Score	FP Neg. Score	SP D. Value	SP Score	SP Neg. Score
Production Share, %	Option 1	0	60	60		30		
	Option 2	0	50			40		
	Option 3	1	40		40	50		50
	Option 4	0	30			55		
	Option 5	0	20			60	60	
B Factor 3	Option 1	0	10	10		6		
	Option 2	0	8			8		
	Option 3	1	6		6	10		10
	Option 4	0	4			12	12	
B Factor 1	Option 1	0	9	9		4		
	Option 2	0	8			6		
	Option 3	0	6			8		
	Option 4	1	4		4	10	10	10
A Factor 1	Option 1	0	6	6		2		
	Option 2	0	5			3		
	Option 3	0	4			4		
	Option 4	0	3			5		
	Option 5	1	2		2	6	6	6
A Factor 2	Option 1	0	5	5		1		
	Option 2	0	4			2		
	Option 3	0	3			3		
	Option 4	1	2		2	4		4
	Option 5	0	1			5	5	
A Factor 3	Option 1	0	5	5		1		
	Option 2	0	4			2		
	Option 3	0	3			3		
	Option 4	0	2			4		
	Option 5	1	1		1	5	5	5

Table 7. *Cont.*

Issue	Possible Options	Optimal Option	FP			SP		
			D. Value	Score	Neg. Score	D. Value	Score	Neg. Score
B Factor 2	Option 1	0	3	3		0		
	Option 2	0	2.5			0		
	Option 3	0	2			1		
	Option 4	1	1		1	2	2	2
A Factor 4	Option 1	1	1	1	1	0	0	0
	Option 2	0	0.5			0		
	Option 3	0	0			0		
	Option 4	0	0			0		
	Option 5	0	0			0		
B Factor 4	Option 1	1	1	1	1	0	0	0
	Option 2	0	0.5			0		
	Option 3	0	0			0		
	Option 4	0	0			0		
			Total Negotiation Value		58	Total Negotiation Value		87

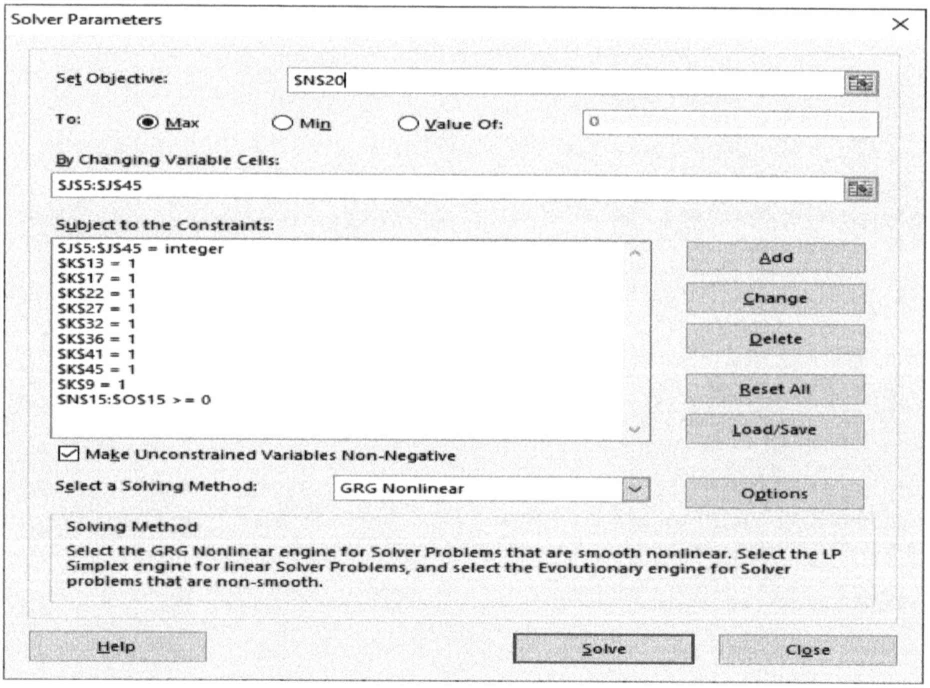

Figure 2. Solver objective and constraints of Nash solution model.

Therefore, we can say that x_{13}, x_{23}, x_{34}, x_{45}, x_{54}, x_{65}, x_{74}, x_{81}, and x_{91} are equal to one, and others are zeros.

Table 8. Solver output of Nash solution of the FP and SP.

Variable	NOC	IOC's
Negotiation Value (1)	58	87
RV (2)	35	65
Excess (3) = (1) − (2)	23	22
Max Feasible (4)	85	100
Potential (5) = (4) − (2)	50	35
POP (6) = (3)/(5)	0.460	0.629
Product (7) = (3) of NOC × (3) of IOC	506	
MinPOP (8) = min ((6) of NOC, (6) of IOC))	0.460	

		Optimal options of the FP and SP								
Solution	Max.	Issue 1	Issue 2	Issue 3	Issue 4	Issue 5	Issue 6	Issue 7	Issue 8	Issue 9
Nash	506	Option 3	Option 3	Option 4	Option 5	option4	option5	option4	option1	Option 1
Variable		Share	B3	B1	A1	A2	A3	B2	A4	B4
Value		15%	0.65	0.98	0.98	0.84	0.75	0.85	0.35	0.40

5.3. Maxi-Min Solution

In the Maxi-min solution, the value of maximizing the minimum POP is improved. The minimum POP is increased from 0.229 to 0.514. This output has been determined with the help of Excel Solver. The formulation and solution to maximize the minimum POP of the negotiation template of the FP and SP are given in Tables 9 and 10.

Table 9. Maxi-min solution of the negotiation score of FP and SP using Solver software.

Issue	Possible Options	Optimal Option	FP			SP		
			D. Value	Score	Neg. Score	D. Value	Score	Neg. Score
Production Share, %	Option 1	0	60	60		30		
	Option 2	0	50			40		
	Option 3	1	40		40	50		50
	Option 4	0	30			55		
	Option 5	0	20			60	60	
B Factor 3	Option 1	1	10	10	10	6		6
	Option 2	0	8			8		
	Option 3	0	6			10		
	Option 4	0	4			12	12	
B Factor 1	Option 1	0	9	9		4		
	Option 2	0	8			6		
	Option 3	0	6			8		
	Option 4	1	4		4	10	10	10
A Factor 1	Option 1	0	6	6		2		
	Option 2	0	5			3		
	Option 3	0	4			4		
	Option 4	1	3		3	5		5
	Option 5	0	2			6	6	
A Factor 2	Option 1	0	5	5		1		
	Option 2	0	4			2		
	Option 3	0	3			3		
	Option 4	0	2			4		
	Option 5	1	1		1	5	5	5
A Factor 3	Option 1	0	5	5		1		
	Option 2	0	4			2		
	Option 3	0	3			3		
	Option 4	0	2			4		
	Option 5	1	1		1	5	5	5

Table 9. *Cont.*

Issue	Possible Options	Optimal Option	FP			SP		
			D. Value	Score	Neg. Score	D. Value	Score	Neg. Score
B Factor 2	Option 1	0	3	3		0		
	Option 2	0	2.5			0		
	Option 3	0	2			1		
	Option 4	1	1		1	2	2	2
A Factor 4	Option 1	1	1	1	1	0	0	0
	Option 2	0	0.5			0		
	Option 3	0	0			0		
	Option 4	0	0			0		
	Option 5	0	0			0		
B Factor 4	Option 1	1	1	1	1	0	0	0
	Option 2	0	0.5			0		
	Option 3	0	0			0		
	Option 4	0	0			0		
			Total Negotiation Value		62	Total Negotiation Value		83

Table 10. Solver output of Maxi-min solution of the FP and SP.

Variable	NOC	IOC's
Negotiation Value (1)	62	83
RV (2)	35	65
Excess (3) = (1) − (2)	27	18
Max Feasible (4)	85	100
Potential (5) = (4) − (2)	50	35
POP (6) = (3)/(5)	0.540	0.514
Product (7) = (3) of NOC × (3) of IOC	486	
MinPOP (8) = min ((6) of NOC, (6) of IOC))	0.514	

Optimal Options of the FP and SP										
Solution	Max	Issue 1	Issue 2	Issue 3	Issue 4	Issue 5	Issue 6	Issue 7	Issue 8	Issue 9
Maxi-min	0.514	Option 3	Option 1	Option 4	Option 4	Option5	option5	option4	option1	Option 1
Variable		Share	B3	B1	A1	A2	A3	B2	A4	B4
Value		15%	0.55	0.98	0.96	0.86	0.75	0.85	0.35	0.40

The negotiation based on the Mix-min solution yields a solution for the FP and SP with the following production share, A factor, and B factor: production share at 15%, B3 at 0.55, B1 at 0.98, A1 at 0.96, A2 at 0.86, A3 at 0.75, B2 at 0.85, A4 at 0.35, and B4 at 0.40, see Table 10.

5.4. Summary of the Effect of the Three Contracts on the Economic Evaluation Model of the AA oil Field

The three outputs of the three agreements, the original EPSA IV, optimized by the Nash solution, and optimized by the Maxi-min solution, are shown in Table 11.

The economic indicators, NPV, and IRR of FP and SP of the three contracts are shown in Table 12 and Figure 3. For the SP, the optimization by using the Nash solution has shown the best improvement. The SP's NPV and IRR are increased from USD 148 million and 15.63% to USD 222 million and 17.94, respectively.

Table 11. Summary of the issues and options of the three contracts of the FP and SP.

Solution	OF *	Issue 1	Issue 2	Issue 3	Issue 4	Issue 5	Issue 6	Issue 7	Issue 8	Issue 9
Variable		Share	B3	B1	A1	A2	A3	B2	A4	B4
Original Agreement		Option 3	Option 2	Option 3	Option 1	Option2	Option2	Option3	Option2	Option 1
Value		15%	0.6	0.95	0.90	0.80	0.60	0.80	0.40	0.40
Nash	506	Option 3	Option 3	Option 4	Option 5	Option 4	Option 5	Option 4	Option1	Option 1
Value		15%	0.65	0.98	0.98	0.84	0.75	0.85	0.35	0.40
Maxi-min	0.514	Option 3	Option 1	Option 4	Option 4	Option5	Option 5	Option 4	Option1	Option 1
Value		15%	0.55	0.98	0.96	0.86	0.75	0.85	0.35	0.40

* Objective Function value.

Table 12. The effect of the original agreement, Maxi-min, and Nash solutions on the economic indicators of FP and SP.

	FP		SP	
Solutions	NPV, USD MM	IRR, %	NPV, USD MM	IRR, %
Original Agreement	5386	409	148	15.63
Nash Solution	5312	406	222	17.94
Maxi-min Solution	5339	408	195	17.01

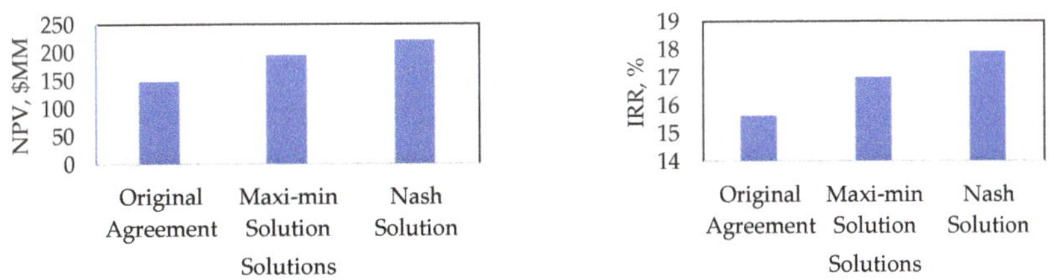

(**a**) NPV, USD MM (**b**) IRR, %

Figure 3. The SP's NPVs and IRRs of the original agreement and the optimization solutions of the Maxi-min and Nash solutions.

The above calculations show the impact of applying the proposed two models on the performance of both parties. The RV for both parties was respected, and better scores were found. Eventually, the effect on NPV and IRR was found to be promising. The effect of different levels of the nine decision variables was found in the literature. Different variables were found to have different weights (Balhasan, et al., 2020). However, determining the best options for the levels based on these weights is new in this study. Decision makers in both parties can utilize the tool used in this study to enhance their agreement terms based on a win-win strategy. Using a common tool can reduce the needed efforts in the negotiation process and reduce the conflict between both parties. The tool used can be easily understood and applied by practitioners. The previous results, especially Figure 3, show how useful it is to use the proposed tool.

6. Conclusions

The EPSA agreement is a complicated method of equity split used in the oil industry. Usually, a production-based sliding scale and R-factor system is used. The SP's NPV and IRR from the original EPSA agreement conditions were USD 148 million and 15.63%, respectively. At the beginning, the IRR was too low to satisfy the SP. Therefore, better configurations were needed. Two approaches were used to find the best negotiation agreement. The Maxi-min solution maximizes the minimum of the two parties' proportion of the POP. The Nash solution maximizes the product of excesses. The two models have shown a significant improvement in the SP's NPV and IRR. The Nash solution has shown the best improvement in favor of the SP. The SP's NPV and IRR were increased from USD 148 million and 15.63% to USD 222 million and 17.94, respectively. The Maxi-min solution also showed an improvement, but less than the Nash solution. The SP's NPV and IRR were increased from USD 148 million and 15.63% to USD 195 million and 17.01, respectively. Such gains for the SP were acceptable by the FP. The two parties achieved agreement optimization.

There are some limitations in this study. For example, the study presents the results for a certain case study. More case studies, especially in the region, can provide more insights. Moreover, Excel Solver does not guarantee an optimal solution always. Other methods for optimization, such as the genetic algorithm, can be investigated in the future. Moreover, the EPSA agreements can contain other negotiation issues that can be investigated in further research.

Author Contributions: Conceptualization S.B. and S.S.; methodology, S.B. and S.S.; software, S.B. and M.A.; validation, M.A., S.B. and M.I.T.; formal analysis, M.A. and S.B.; investigation, S.B.; resources, B.S., S.S., M.I.T. and S.B.; data curation, S.B.; writing—original draft preparation, S.B. and M.A.; writing—review and editing, S.B., S.S., M.I.T. and M.A.; visualization, B.S., W.S. and S.S.; supervision S.S. and S.B.; project administration, B.S., W.S. and S.B.; funding acquisition, B.S. and W.S. All authors have read and agreed to the published version of the manuscript.

Funding: This study received funding from King Saud University, Saudi Arabia through researchers supporting project number (RSP-2021/145). Additionally, the APCs were funded by King Saud University, Saudi Arabia through researchers supporting project number (RSP-2021/145).

Data Availability Statement: Not applicable.

Acknowledgments: The authors would like to thank King Saud University, Riyadh, Saudi Arabia, with researchers supporting project number RSP-2021/145. Furthermore, the authors would like to thank the Libyan National Corporation for providing the necessary data for this paper. Thanks in particular to A. Omar, A. Ali, and O. Mohamed for their assistance and assistance throughout the course of this work.

Conflicts of Interest: The authors declare no conflict of interest.

Nomenclature

FP	First Party
SP	Second Party
NPV	Net Present Value
IRR	Internal Rate of Return
EPSA	Exploration and Production Sharing Agreement
GDP	Gross Domestic Product
POP	Proportional of the Potential
RV	The Reservation Value
CAPEX	Capital Expenditures
OPEX	Operating Costs
LHP	liquefied hydrocarbon by products
NCF	Net Cash Flow
EOR	enhanced oil recovery
NOC	National Oil Corporation
IOC	International Oil Company

References

1. Kemper, R.E.; Kemper, D.R. *Negotiation Literature*; Scarecrow Press: Metuchen, NJ, USA, 1994.
2. Raiffa, H. *The Art and Science of Negotiation*; Belknap: Cambridge, MA, USA, 1982.
3. Pruitt, D.G.; Carnevale, P.J. *Negotiation in Social Conflict*; Brooks/Cole Pub. Co.: Pacific Grove, CA, USA, 1993.
4. Yas, H. Economic Evaluation of Petroleum Projects (Basic Aspects). *J. Pet. Res. Stud.* **2010**, *1*, 37–42. [CrossRef]

5. Mian, M.A. *Project Economics and Decision Analysis Volume 1: Deterministic Models*; PennWell Publishing Co.: Tulsa, OK, USA, 2011; pp. 73–81.
6. Johnston, D. *International Exploration Economics. Risk, and Contract Analysis*; PennWell: Tulsa, OK, USA, 2003.
7. Balhasan, S.A.; Towler, F.B.; Miskimins, J.L. Modifying the Libyan Fiscal Regime to Optimize its Oil Reserves and Attract More Foreign Capital Part 1: LEPSA I Proposal. *OPEC Energy Rev. J.* **2012**, *37*, 220–246. [CrossRef]
8. Balhasan, S.A.; Towler, F.B.; Miskimins, J.L. Modifying the Libyan Fiscal Regime to Optimize its Oil Reserves and Attract More Foreign Capital Part 2: Application to CO2-EOR Project in Libya. *OPEC Energy Rev. J.* **2012**, *37*, 270–313. [CrossRef]
9. Demirmen, F. Win-Win Upstream Fiscal Systems: What They Are and How to Achieve Them. In Proceedings of the SPE Hydrocarbon Economics and Evaluation Symposium, Dallas, TX, USA, 8–9 March 2010.
10. Rutledge, I.; Wright, P. Profitability and taxation: Analyzing the distribution of rewards between company and country. *Energy Policy* **1998**, *26*, 795–812. [CrossRef]
11. Stermole, F.J.; Stermole, J.M. *Economic Evaluation and Investment Decision Methods*; Investment Evaluation Co.: Lakewood, CA, USA, 2006; pp. 223–234.
12. Bratvold, R.B.; Koch, F. Game Theory in the Oil and Gas Industry. *Way Ahead* **2011**, *7*, 18–20. [CrossRef]
13. Ibrahim, M.A.R.; Jaini, N.I.; Khalif, K.M.N.K. A Review of Game Theory and Multi-Criteria Decision-Making Methods with 10 Application to the Oil Production and Price. *Malays. J. Appl. Sci.* **2022**, *7*, 10–26.
14. Boyce, J.R.; Vojtassak, L. An 'oil'igopoly theory of exploration. *Resour. Energy Econ.* **2008**, *30*, 428–454. [CrossRef]
15. Willigers, B.J.; Hausken, K.; Bratvold, R. Uncertainty and preferences in a joint E&P development program analyzed in a game-theoretic framework. *J. Pet. Sci. Eng.* **2010**, *74*, 88–98.
16. Esmaeili, M.; Bahrini, A.; Shayanrad, S. Using game theory approach to interpret stable policies for Iran's oil and gas common resources conflicts with Iraq and Qatar. *J. Ind. Eng. Int.* **2015**, *11*, 543–554. [CrossRef]
17. Langer, L.; Huppmann, D.; Holz, F. Lifting the US crude oil export ban: A numerical partial equilibrium analysis. *Energy Policy* **2016**, *97*, 258–266. [CrossRef]
18. Tominac, P.; Mahalec, V. A game theoretic framework for petroleum refinery strategic production planning. *AIChE J.* **2017**, *63*, 2751–2763. [CrossRef]
19. Moradinasab, N.; Amin-Naseri, M.R.; Behbahani, T.J.; Jafarzadeh, H. Competition and cooperation between supply chains in multi-objective petroleum green supply chain: A game theoretic approach. *J. Clean. Prod.* **2018**, *170*, 818–841. [CrossRef]
20. Araujo, F.C.; Leoneti, A.B. Game Theory and 2 × 2 Strategic Games Applied for Modeling Oil and Gas Industry Decision-Making Problems. *Pesqui. Oper.* **2018**, *38*, 479–497. [CrossRef]
21. Nicoletti, J.; You, F. Multiobjective economic and environmental optimization of global crude oil purchase and sale planning with noncooperative stakeholders. *Appl. Energy* **2020**, *259*, 114222. [CrossRef]
22. Babaei, M.; Asgarian, F.; Jamali, M.B.; Rasti-Barzoki, M.; Piran, M.J. A game theoretic approach for pricing petroleum and determining investors' production volume with the consideration of government and intermediate producers. *Sustain. Energy Technol. Assess.* **2020**, *42*, 100825. [CrossRef]
23. Xue, Q.; Cheng, C.; Wang, Z.; Liu, M.M. Bargaining strategy of oil companies in international oil and gas development Projects–Based on a bilateral bargaining model. *Pet. Sci.* **2021**, *18*, 1270–1279. [CrossRef]
24. Araujo, F.C.; Leoneti, A.B. Evaluating the stability of the oil and gas exploration and production regulatory framework in Brazil. *Group Decis. Negot.* **2020**, *29*, 143–156. [CrossRef]
25. Csercsik, D. A potential mechanism of gas supply-security cooperation based on a game-theoretic model. *arXiv* **2022**, arXiv:2209.05089.
26. Keshavarz, M.; Iranmanesh, H.; Dehghan, R. Modelling the Iranian Petroleum Contract fiscal regime using bargaining game theory to guide contract negotiators. *Pet. Sci.* **2021**, *18*, 1887–1898. [CrossRef]
27. Dirani, F.; Ponomarenko, T. Contractual systems in the oil and gas sector: Current status and development. *Energies* **2021**, *14*, 5497. [CrossRef]
28. Balhasan SMisbah, B.; Omar, M.; Musbah, I.; Khameiss, B. Impact of Proposal Changes to Libyan Oil Taxation System on Developing the 137 B Offshore Field. *J. Eng. Appl. Sci.* **2018**, *15*, 6085–6090.
29. Balhasan, S.A.; Towler, B.F.; AlHadramy, K.O.; Biltayib, B.M. Comparative risk evaluation and sensitivity analysis of the Libyan EPSA IV and its modified model LEPSA I. *Petroleum* **2020**, *6*, 206–213. [CrossRef]
30. Nash, J.F. The bargaining problem. *Econom. J. Econom. Soc.* **1950**, *18*, 155–162. [CrossRef]
31. Okoro, J.O.; Oghenekevwe, A.; Oladejo, N.K. The game theory and simulation with finite mixed strategy of two-players zero sum game, and. *Asian J. Math. Comput. Res.* **2018**, *24*, 218–233.
32. Turbay, G.; Reyes, G.E. On the Solutions of Games in Normal Forms: Particular Models based on Nash Equilibrium Theory. *Mediterr. J. Soc. Sci.* **2019**, *10*, 1–7. [CrossRef]
33. Robinson, D.; Goforth, D. *The Topology of the 2 × 2 Games: A New Periodic Table*; Routledge: New York, NY, USA, 2005.
34. Balhasan, S.; Biltayib, B. Risk Evaluation & Sensitivity Analysis of the Exploration & Production Sharing Agreement. In Proceedings of the Abu Dhabi International Petroleum Exhibition & Conference, Abu Dhabi, United Arab Emirates, 11–14 November 2019. SPE-197442-MS.

energies

MDPI

Article

Market Integration, Industrial Structure, and Carbon Emissions: Evidence from China

Kun Zheng [1], Hongbing Deng [1,*], Kangni Lyu [1], Shuang Yang [2,*] and Yu Cao [3,*]

1 School of Economics and Management, China University of Geosciences, Wuhan 430078, China
2 Department of Information Engineering, Wuhan Marine College, Wuhan 430062, China
3 Business School, Xinyang Normal University, Xinyang 464000, China
* Correspondence: deng115@cug.edu.cn (H.D.); 20151002496@cug.edu.cn (S.Y.); xynucaoyu@xynu.edu.cn (Y.C.)

Abstract: Against the backdrop of China's carbon emission reduction targets and the promotion of the construction of a unified domestic market, what kind of carbon emission effect has market integration had in weakening regional barriers and optimizing resource allocation? This paper adopts a two-way fixed effects analysis based on China's provincial panel data from 2003 to 2019. It uses a mediation model to explore the relationship between market integration and carbon emissions. Furthermore, industrial rationalization and upgrade are the basis for examining whether a competitive or cooperative relationship exists between the carbon emission effects generated and promoting market integration between regions. The study finds a negative relationship between market integration and carbon emissions. In addition, there is significant heterogeneity in the effect of market integration on carbon emissions, and the influence effect is mainly in central China; industrial rationalization can play an enhanced role in the process of the negative impact of market integration on carbon emissions, further enhancing the negative contribution of market integration to carbon emissions. However, market integration can weaken its negative impact on carbon emissions with the industrial upgrade, and there may still be invisible barriers between regions in promoting market integration barriers.

Keywords: market integration; carbon emissions; industrial rationalization; industrial upgrade

Citation: Zheng, K.; Deng, H.; Lyu, K.; Yang, S.; Cao, Y. Market Integration, Industrial Structure, and Carbon Emissions: Evidence from China. *Energies* **2022**, *15*, 9371. https://doi.org/10.3390/en15249371

Academic Editor: Charisios Achillas

Received: 8 November 2022
Accepted: 8 December 2022
Published: 11 December 2022

1. Introduction

Market integration aims to eliminate barriers to the flow of resources and factors in each region, break down administrative and trade barriers, form a standardized and orderly market resource-sharing and cooperation platform, and promote equal cooperation and fair competition in economic development between regions. Since the reform and opening up, along with the enhanced flow of resources and factors between regions, the economic development of China has become more and more closely linked, and market integration has become an essential path for economic growth. With the introduction of the carbon peak and neutrality targets, all regions must consider environmental factors in the process of economic development, and try to achieve "zero emissions" in the development process. China has issued guidelines on establishing a unified domestic market, indicating that the unification of factor and resource markets is inevitable, and how to take into account the two-way results of market expansion and carbon emission reduction deserves attention and in-depth analysis.

For a long time, due to the effects of inter-regional competition, regional barriers have deepened, and industrial isomorphism has become increasingly aggravated. This has led to massive energy consumption and exacerbated the rise of carbon emissions, severely hindering green environmental development. The deepening of regional development strategy has produced some policy promotion effects on breaking regional barriers. The guidelines on establishing a unified domestic market have laid a solid foundation for further deepening market integration, and have contributed to the positive environmental effects generated in the process of market integration. Therefore, bringing industrial structure and agglomeration into the framework of discussion, this paper discusses the impact mechanism and effect of market integration on carbon emission levels, and then explores the path and policies to optimize carbon emissions around the construction of market integration.

2. Literature Reviews and Theoretical Analysis

2.1. Literature Reviews

The ability of regional integration strategies to weaken market segmentation and reduce carbon emissions is crucial to achieving China's carbon emission reduction targets. However, the current literature concerning the impact of market integration on carbon emissions focuses more on its effects on environmental quality, and only a few pay attention to the impact on carbon emissions.

In studies on the impact of trade openness on environmental quality, market integration leads to more unrestricted output flows, implying the elimination of trade barriers. However, reducing trade barriers has a nonlinear effect on the environment [1–4]. There are two explanations for the nonlinear relationship between trade openness and the environment. One uses the well-known pollution haven hypothesis (PHH) to explain the nonlinear relationship. PHH argues that since pollution often occurs during the transfer of industries from developed to developing countries, low-income developing countries are always the victims of environmental pollution. The environmental impact of trade openness is detrimental to developing countries but benign to developed countries; thus, global environmental performance is conditional [5–8]. However, trade integration, while increasing the intensity of pollutant emissions, has offset this negative effect by increasing efficiency and promoting cleanliness in its manufacturing sector [9]. Therefore, in the context of trade openness, PHH is often used to test the environmental impact of increased foreign direct investment (FDI). The other explanation has argued that the results of trade openness can be divided into scale, technology, and composition effects [7,10–12]. The scale effect refers to the expansion of pollutant emissions as the size of the economy increases due to trade liberalization. The technology effect refers to the upgrading of green technologies to reduce the intensity of emissions, and through stricter environmental regulation, trade openness will raise income levels and increase the demand for a cleaner environment. The composition effect refers to the two-way impact of industrial restructuring. If the high-polluting industrial sector has a comparative advantage, trade openness can damage the environment by making local areas more specialized in high-polluting production. Otherwise, trade openness can improve the environment by making local areas more specialized in cleaner production [2,3,11]. Previous literature has focused extensively on the environmental impacts of trade barriers. However, most have focused on the ecological effects of international trade, while relevant evidence from domestic trade remains scarce.

For studies on the impact of market integration on environmental quality, Li and Lin evaluated the carbon emission performance of provincial regions in China from 1995 to 2012 using a non-radial directional distance function to investigate the effect of regional market integration on carbon emission performance. The study found that regional market integration can significantly contribute to carbon emissions [13]. He et al. confirmed the significant contribution of regional market integration to the marginal abatement cost of carbon emissions in 30 Chinese provinces during 2002–2011 [14]. Lin and Du (2015) used a Tobit regression model to estimate the effect of market-oriented reforms on carbon emissions efficiency [15]. The results show that market-oriented reforms can improve carbon emissions efficiency. The above studies all consider the relationship between regional market integration and carbon emissions. In addition, these studies also proved that the increase in the level of market integration could meaningfully contribute to the strengthening of regional market forces, especially in the area of high-carbon markets, enhancing the competitiveness of enterprises and investment development opportunities [16]. However, it is worth noting that to improve carbon productivity, it is necessary to strengthen inter-regional cooperation further and focus on the coordinated development of carbon productivity in the development process [17]. Certainly, some scholars focus more on the environmental impact of market integration on energy and electricity. They believe the integration process will increase energy and electricity consumption and play a vital role in developing renewable energy. Within the scope of a unified market, technological progress and the strengthening of environmental regulation tends to promote renewable energy development, thereby accelerating the process of sustainable development [18]. On a larger scale, such as regional energy market cooperation that transcends national borders, it faces many additional problems [19].

In summary, market integration may affect carbon emissions. Market integration is accompanied by the free flow of production factors, which involves economic and technological innovation and carbon emissions [20,21]. However, only a few studies focus on the relationship between market integration and carbon emissions. Similar to other potential influencing factors of carbon emissions, other external conditions may influence the effect of market integration on carbon emissions. The impact of market integration on carbon emissions may vary depending on the external environment.

To address some of these possible gaps in existing studies, this paper verifies the relationship between market integration and carbon emissions from the perspective of domestic trade barriers, using provincial panel data for 30 provinces (excluding Tibet) from 2003 to 2019. The paper aims to contribute to an understanding of the relationship between market integration policies to reduce regional development inequalities and carbon emissions in China and other developing countries.

The potential contributions of this paper can be divided into three aspects. First, we explore the possible negative effects of market integration on carbon emissions in Chinese provinces and cities. Second, unlike most studies examining the environmental impact of international trade, we provide new empirical evidence from the perspective of enhanced factor mobility. Third, we argue that there is a significant correlation between market integration and carbon emissions, with a mediating effect through changes in industrial structure.

2.2. Theoretical Analysis and Research Hypothesis

Market segmentation leads to resource misallocation, which results in the inability to achieve free flow of factor resources within a region, making it challenging to allocate regional resources efficiently, and then adversely affecting the carbon emission intensity in the long run [22]. When local governments engage in integrated cooperation, the level of inter-regional market integration gradually increases, and factors of production can realize a free flow, which will significantly promote the energy-saving and emission-reduction effects of urban agglomerations [23]. Market integration refers to the free flow of goods and factors of production within a framework of consistent rules between regions and industrial sectors, which will lead to scale economy, knowledge sharing, and technology spillovers [24]. In other words, market integration can indirectly affect carbon emissions through economic growth and technological progress [5,25–27]. This paper analyzes the impact of market integration on carbon emissions, focusing on the possible scale effects, structural effects, and regional heterogeneity.

Market integration may increase environmental by-products through the expansion of local markets and the promotion of enterprise production, which exacerbates environmental pollution, reflecting the scale effect of market integration. Generally speaking, market integration implies free trade of commodities and removing barriers to factor flow, which is conducive to optimizing the economic structure and developing regional scale effects, thus improving resource allocation efficiency and production technology progress and promoting pollution reduction. While the expansion of trade and market scale caused by market integration may aggravate carbon emissions due to increased production, but also reduce pollution emissions due to the scale effect at the same time, the conclusion depends on comparing different forces [1].

Based on the above theoretical analysis, hypothesis 1 is proposed: Market integration will promote the scale effect and increase production quantity, thus leading to increased carbon emissions.

The environmental Kuznets curve theory suggests that market integration reduces environmental pollution through factors such as industrial agglomeration and industrial restructuring [1]. In the classical new economic geography model, market integration promotes the realization of industrial agglomeration, contributes to the completion of enterprise agglomeration externalities, realizes the diffusion and sharing of environmental protection technology, and contributes to the improvement of green growth efficiency [28–30]. In addition, industrial transfer is often accompanied by policy orientation, while industrial restructuring and development in the region are often closely linked to environmental policies, especially for the transfer of heavily polluting industries, and the effect of such policy suppression is evident [21]. Therefore, it is likely to result in competition between regions in policy development and implementation, which is not conducive to the synergy of industries in each region to reduce carbon emissions. Market integration promotes the transformation and upgrading of industrial structure, and reduces the carbon emission intensity of enterprise production [31–34].

Based on the above theoretical analysis, hypothesis 2 is proposed: Market integration will slow down the carbon emissions increase by influencing industrial restructuring.

The improvement of environmental welfare by market integration also relies on the spatial spillover properties of pollutants and cross-regional pollution coefficients. In particular, with the changing focus of China's regional development strategy and the accelerated reform process [35], different regions have different economic development statuses, degrees of market integration, and energy structures. The impact of their market integration on carbon emission levels also varies [36].

Based on the above theoretical analysis, Hypothesis 3 is proposed: Regional heterogeneity in the impact of market integration on carbon emission levels.

The remainder of the paper is as follows: Section 3 details the model construction and data description; Section 4 presents and discusses the regression results and conducts robustness tests; Section 5 presents the main conclusions and policy implications.

3. Data and Method

3.1. Data Description

3.1.1. Carbon Emissions Measurement

This paper takes carbon emissions as a dependent variable, and the data comes from China Emission Accounts and Datasets: https://www.ceads.net (accessed on 1 August 2022). It includes carbon emissions from both fossil fuel combustion (i.e., energy-related emissions) and cement production (process-related emissions) in the emission accounts. Energy-related carbon emissions are converted from the carbon content in fossil fuels. We use mass balances to calculate emissions according to the IPCC guidelines (2006), the formula is as follows:

$$CE_i = AD_i \times NCV_i \times CC_i \times O \quad (1)$$

In the equation, CE_i refers to carbon emissions from fossil $fuel_i$. While China's statistical energy system has 26 types of fossil fuels, references to the calculation formula and method of carbon emission by existing scholars [37,38], merge them into 17 types due to the small consumption amount and similar quality of certain fuels. Ad_i is the "activity data" used for emission estimation. In the case of energy-related emission accounting, AD_i refers to the combustion volume of fossil fuel i. NCV_i represents the "net caloric value," which is the heat value per physical unit from the combustion of fossil fuel i. CC_i is the "carbon content" of fuel i, quantifying carbon emissions per net caloric value produced. O refers to "oxygenation efficiency," which represents the oxidation ratio during fossil fuel combustion of certain fuels.

By aggregating emission results from different energy types, the formula for calculating the total carbon emissions of a province is as follows:

$$TCE = \sum CE_i \tag{2}$$

3.1.2. Market Integration Measurement

We use the relative price method proposed by Parsley, Wei and Poncet to measure market integration [26,39]. First, calculate the variance using the absolute value of the relative price of the commodity $\left|\Delta Q^k_{ijt}\right|$. The formula is as follows:

$$\Delta Q^k_{ijt} = \ln\left(P^k_{it}/P^k_{jt}\right) - \ln\left(P^k_{it-1}/P^k_{jt-1}\right) \tag{3}$$

By simply morphing, ΔQ^k_{ijt} can be expressed as a chain index of commodity prices P^k_{it}/P^k_{jt} and P^k_{it-1}/P^k_{jt-1}. The formula is as follows:

$$\Delta Q^k_{ijt} = \ln\left(P^k_{it}/P^k_{it-1}\right) - \ln\left(P^k_{jt}/P^k_{jt-1}\right) \tag{4}$$

To calculate and measure the market segmentation level more accurately and reflect its actual situation, therefore, in a further transformation of the equation, the non-additive effects due to commodity heterogeneity $\left|\Delta Q^k_{ijt}\right|$ are first hypothesized by means using the mean value method:

$$\left|\Delta Q^k_{ijt}\right| = \alpha^k + \varepsilon^k_{it} \tag{5}$$

To eliminate α^k, we need to find the mean value of $\left|\Delta Q^k_{ijt}\right|$, and subtract that mean by $\left|\Delta Q^k_{ijt}\right|$. The difference between the two is obtained as q^k_{ij}. After that, the factors affecting q^k_{ij} are mainly focused on the market score and some random factors. Calculate the variance $Var\left(q^k_{ij}\right)$ using the q^k_{ij} values. Variance $Var\left(q^k_{ij}\right)$ indicates the change in the arbitrage space due to market segmentation. If this arbitrage space is smaller, it shows a higher level of current market integration and vice versa. The mean value of $Var\left(q^k_{ij}\right)$ of a province and city and all bordering provinces and cities can be used to express the degree of market segmentation of this province and city, that is:

$$Var\left(q^k_m\right) = \left(\sum_{i \neq j} Var\left(q^k_{ij}\right)\right)/N \tag{6}$$

Among them, j denotes all provinces and cities bordering province i, m represents the name of the province and city, and N represents the number of combinations of provinces and cities bordering province and city i. Finally, the market integration index is built on top of the existing segmentation index (expressed as MI, m still indicates the name of the province or city), and the formula for the integration index is defined as follows:

$$MI_m = \sqrt{\frac{1}{Var^k_m}} \tag{7}$$

Therefore, the relationship between the two indices of market segmentation and market integration has an inverse trend. After calculating each provincial and municipal integration index, the average value of all provincial and municipal market integration values in the region can be calculated to measure the market integration level of a particular region. We selected eight categories of: food, tobacco and alcohol; clothing; housing; household goods and services; transportation and communication; education; culture and entertainment; and health care. These eight categories of consumer price index are measured; all data are from provincial statistical yearbooks and the National Bureau of Statistics.

3.1.3. Industry Change Measurement

Industrial rationalization. We refer to existing scholarly practice to measure [40]. The first step is to calculate the structural deviation factor:

$$\mathrm{INDR}^* = \sum_j \frac{Y_j}{Y}\left|\frac{Y_j/L_j}{Y/L} - 1\right| (j = 1,2,3) \tag{8}$$

where i represents each province, j = 1,2,3 represents the three industries respectively, Y_j represents the value added of the industry in that year, and L_j represents the number of employees in that year. INDR* is the degree of industrial structure rationalization, measured by the degree of structural deviation in province i. The summed numbers indicate the relative degree of imbalance between the respective value-added shares of the three industries and the employment shares. In this case, the higher the value of INDR^*, the lower the degree of industrial structure rationalization of the province. Since all data in the index are from the same year, we omit the time subscript t for symbolic simplicity, so as to not cause conceptual confusion.

In the second step, a numerical extreme difference transformation is used to normalize the range of indicators to a specific interval and convert them into positive indicators.

$$\mathrm{INDRM}_i = \frac{\underset{k}{\mathrm{maxmind}}_k^* - \mathrm{indrL}_i^* + 1}{\underset{k}{\max}\,\mathrm{indr}_k^* - \mathrm{minindr}_k^*}\alpha \tag{9}$$

In the third step, we first use the structural deviation to calculate the degree of rationality of the industrial structure, which is given by:

$$\mathrm{INDRS} = \sum_{i=1}^{n}\left|\frac{Y_i/L_i}{Y/L} - 1\right| = \sum_{i=1}^{n}\left|\frac{Y_i/Y}{L_i/L} - 1\right|^{L} \tag{10}$$

In this equation, Thiel's index is introduced, and the formula is as follows:

$$\mathrm{INDR} = \sum_{i=1}^{n}\left(\frac{Y_i}{Y}\right)\ln\left(\frac{Y_i}{L_i}/\frac{Y}{L}\right) \tag{11}$$

where, INDR denotes the Thiel index, which indicates the level of industrial structure upgrading. y denotes output, L denotes employment, i denotes industry, and n denotes the number of industrial sectors; Y/L indicates productivity, Y_i/Y denotes the output structure, and L_i/L indicates the employment structure, and the value range of INDR is (0,LnN). When INDR = 0, it means that the industrial development is very reasonable, and the smaller the INDH value is, the more reasonable the industrial structure is, and the development of each internal factor is balanced.

Industrial upgrade. The upgrading of industries involves the evolution of proportional relations and the improvement of labor productivity. When the share of industries with higher labor productivity in a country or region is more prominent, it indicates a stronger industrial chain heightening in that region. Therefore, following the approach of Liu Wei et al. [41], the connotation of industry chain heightening (INDH) is defined as the weighted value of the product of the proportional relationship between industries and the labor productivity of each industry. This shows the essential characteristics of the evolution of industry chains as higher proportions of labor productivity. The specific formula for measuring the quality of industry chain heightening is:

$$\mathrm{INDH}_{it} = \sum_{m=1}^{3}\frac{Y_{itm}}{Y_{it}} \times \frac{Y_{itm}}{L_{itm}}, m = 1,2,3 \tag{12}$$

Here, Y_{itm} denotes the value added of industry m in period t in region i, L_{itm} represents the number of people employed in industry m in period t in region i. Y_{itm}/Y_{it} denotes the labor productivity of industry m in period t in region i. Considering that labor productivity has a quantitative dimension, this paper adopts the mean value method for dimensionless treatment. All data are from provincial statistical yearbooks, the National Bureau of Statistics and CSMAR Database.

Other control variables. To control external factors in different regions, government input (GOVI), the level of openness to the outside world (FDI), the level of technological innovation (ZLSP), and the level of regional economic development (GDP) are selected as control variables. Government input, the level of openness to the outside world, the level of technological innovation, and the level of regional economic development not only affect carbon emissions [42,43] but also affect the impact of market integration on carbon emissions [44,45].

Government input (GOVI) is measured by the annual fiscal revenue of each province; the level of openness to the outside world (FDI) is measured by the sum of total annual import and export and foreign investment in each province; the level of technological innovation (ZLSP) is measured by the annual number of patents granted in each province, and the level of regional economic development (GDP) is measured by the annual gross product of each province. The relevant data were logarithmically processed. All data are from provincial statistical yearbooks.

3.2. *Methods*

3.2.1. Two-Way Fixed Effects Model

This paper's two-way fixed effects model was constructed to investigate the linear relationship between market integration and carbon emissions adopting panel data from 2003 to 2019. Considering the estimated coefficients of the double logarithmic, market integration on the carbon emissions equation can be treated as the elasticities of the dependent variables [46]. To eliminate heteroscedasticity effectively, concerning the independent variables, we used TCE to represent the level of carbon emissions, MI is the level of market integration, and X represents the other control variables. We conducted a double logarithmic function as shown below:

$$\mathrm{TCE}_{it} = \alpha_0 + \beta_1 \mathrm{MI}_{it} + \beta_2 \mathrm{X}_{it} + \mu_{it} + \varepsilon_{it} \quad (13)$$

3.2.2. Analysis of the Mechanism

In this paper, a mechanism analysis model was constructed, and based on the benchmark model, indicators related to industrial structure change were added to test the mechanism of action between market integration, industrial structure change, and carbon emissions, and further analyze the impact of market integration on carbon emissions. The model is set as follows.

$$\mathrm{TCE}_{it} = \alpha_0 + \beta_1 \mathrm{MI}_{it} + +\beta_2 \mathrm{INDR}_{it} + \beta_3 \mathrm{Interact1}_{it} + \beta_4 \mathrm{X}_{it} + \mu_{it} + \varepsilon_{it} \quad (14)$$

$$\mathrm{TCE}_{it} = \alpha_0 + \beta_1 \mathrm{MI}_{it} + \beta_2 \mathrm{INDH}_{it} + \beta_3 \mathrm{Interact2}_{it} + \beta_4 \mathrm{lnX}_{it} + \mu_{it} + \varepsilon_{it} \quad (15)$$

where TCE is the carbon emission level, MI is the market integration level, INDR is the level of industrial rationalization level, INDH is the level of industrial upgrade, Interact1 is the interaction term between market integration (MI) and industrial rationalization (INDH), Interact2 is the interaction term between market integration (MI) and industrial upgrade (INDH), and X represents other control variables.

3.2.3. Intermediary Effect Model

We constructed a mediating effect model to test the relationship between market integration, industrial structure change, and carbon emissions. The impact of market integration development on carbon emissions was quantitatively analyzed. Then we tested whether industrial structure change mediates the market integration process affecting carbon emissions. In the next step, the extent of the mediating impact was studied under the premise that there is a mediating effect, and industrial rationalization and upgrade were taken as mediating variables. In summary, the empirical panel data regression model constructed in the article is as follows:

$$TCE_{it} = \alpha_0 + \beta_1 MI_{it} + \beta_3 X_{it} + \mu_{it} + \varepsilon_{it} \tag{16}$$

$$Medi_{it} = \alpha_0 + \beta_1 MI_{it} + +\beta_3 X_{it} + \mu_{it} + \varepsilon_{it} \tag{17}$$

$$TCE_{it} = \alpha_0 + \beta_1 MI_{it} + \beta_2 Medi_{it} + \beta_3 X_{it} + \mu_{it} + \varepsilon_{it} \tag{18}$$

4. Result and Discussion

4.1. Spatial and Temporal Trends

4.1.1. Spatial and Temporal Trends in Market Integration

Through the measurement of market integration levels, it is found that each region's overall market integration level shows a fluctuating upward trend, and most regions have the phenomenon of rising and then falling in Figure 1, such as Beijing, Tianjin, Inner Mongolia, etc. During the sample examination period, most regions reached a peak market integration level in 2016. They then fluctuated, indicating to a certain extent that the inter-regional market integration promotes factor flow, and that there is a limited value in the collaborative division of labor. In the market integration process, especially after the local benefits or industrial system have formed a stable situation with a particular economic foundation, most regions focused more on maximizing the local economy, thus gradually lowering the goal of inter-regional cooperation with the risk that market segmentation will rise again. Some regions, such as Shanghai, Zhejiang, and Guangdong, gradually rebounded after showing a downward trend in 2016, and the degree of inter-regional regional cooperation was further strengthened. The level of market integration in the region was continuously enhanced after 2016.

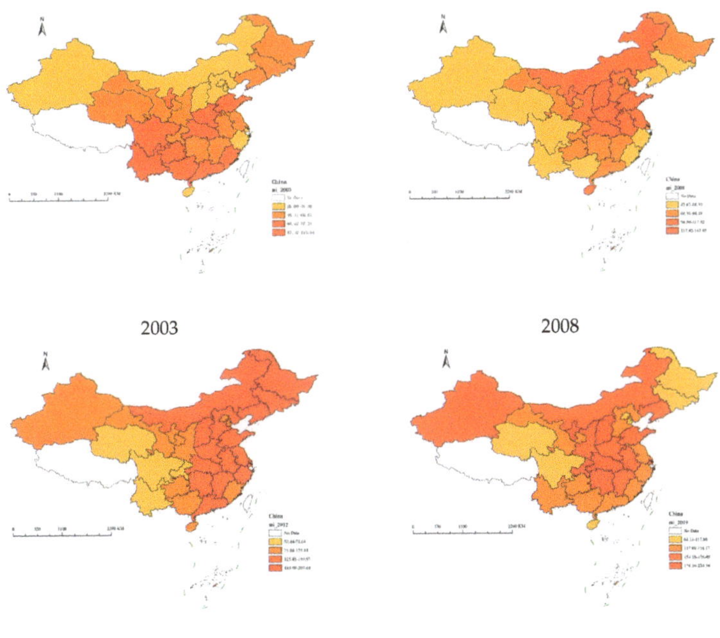

2003 2008

2012 2019

Figure 1. The level of market integration in China.

From a sub-regional perspective, the overall difference in market integration levels among the eastern, central, western, and northeastern regions at the initial stage is slight in Figure 2, the eastern pre-coastal regions are far more demanding than other regions in China in terms of opening and cooperation, are more dependent on industrial chain development and foreign trade cooperation, and are more in need of a market. It is necessary to pursue the role of a win-win or leading role in the process of market integration. On the other hand, after 2012, with the central government's slogan "lucid waters and lush mountains are invaluable assets," the demand for green economic development in this region is increasing, indirectly promoting inter-regional market-level integration. Although the eastern region maintained an upward trend until 2012, it rose slowly, and unlike other areas, it showed a downward trend after 2012 and rebounded in 2016. Still, the gap in the level of market integration in 2019 compared with the central and western regions is evident. It may be because the eastern region, as the frontier region of China's economic development, has shown a high level of inter-regional cooperation and development in the early stage of development and tends to converge in the market integration process. With the central government proposing industrial transformation and structural adjustment, the eastern region may have been in a temporary period of pain since 2012, and it will recover the level of market integration in due course.

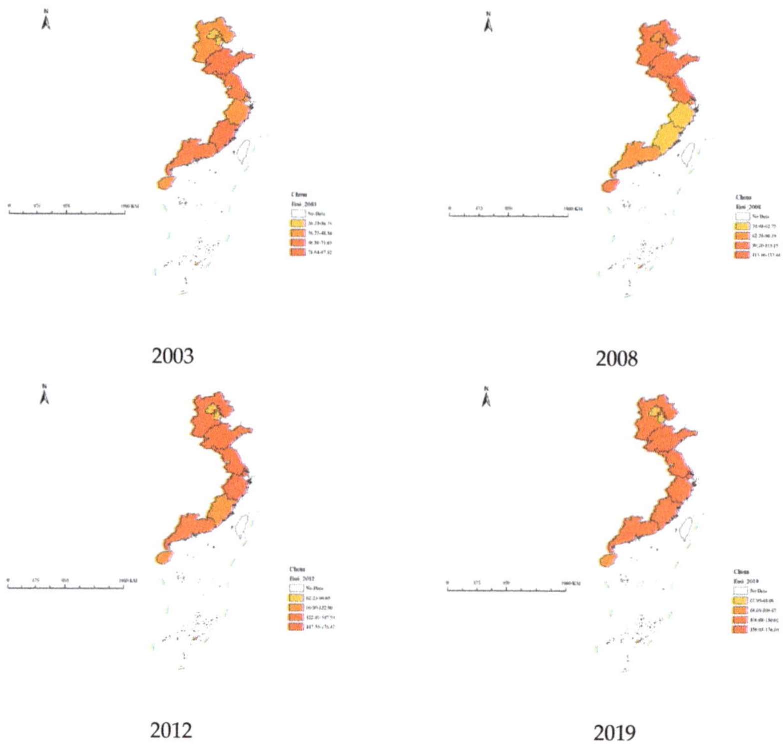

Figure 2. The level of market integration in eastern China.

The central region always showing an upward trend and maintaining a leading position until 2009 in Figure 3, which may also be closely related to the strategic development of the rise of central China that the Chinese government has advocated.

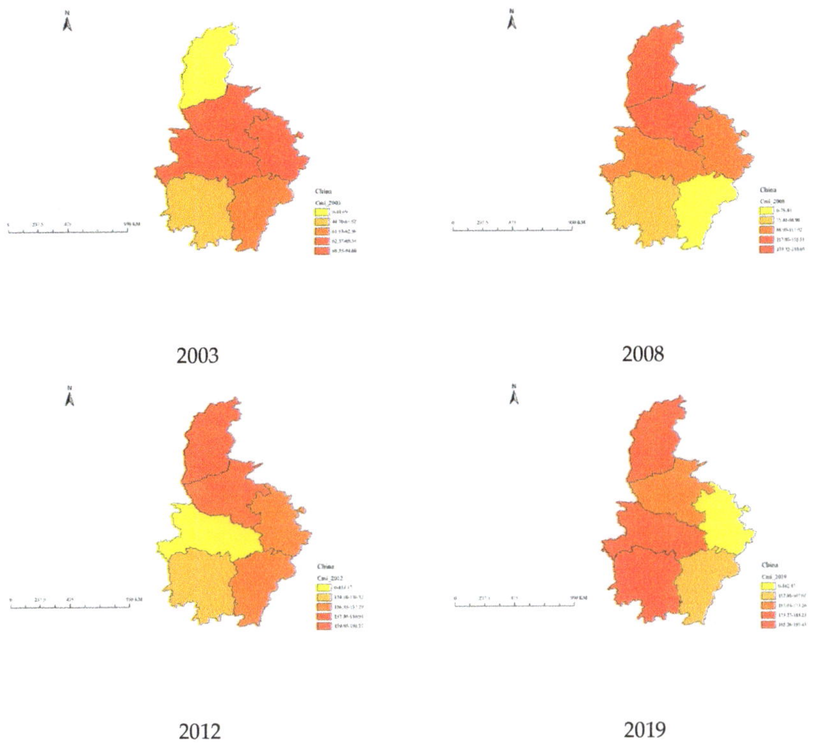

Figure 3. The level of market integration in central China.

The level of market integration in the western region has always been on the rise in Figure 4, showing a slower growth between 2003 and 2012, but between 2012 and 2016, the level of market integration in the western region achieved a significant increase, which may be attributed to the enhanced binding of resources and environment. The western region must seek a path that fits ecological and environmental protection with economic development, so the past development model that relied on environmental resources needs to be improved to rapidly promote the transformation of the region's economic structure through effective inter-regional cooperation.

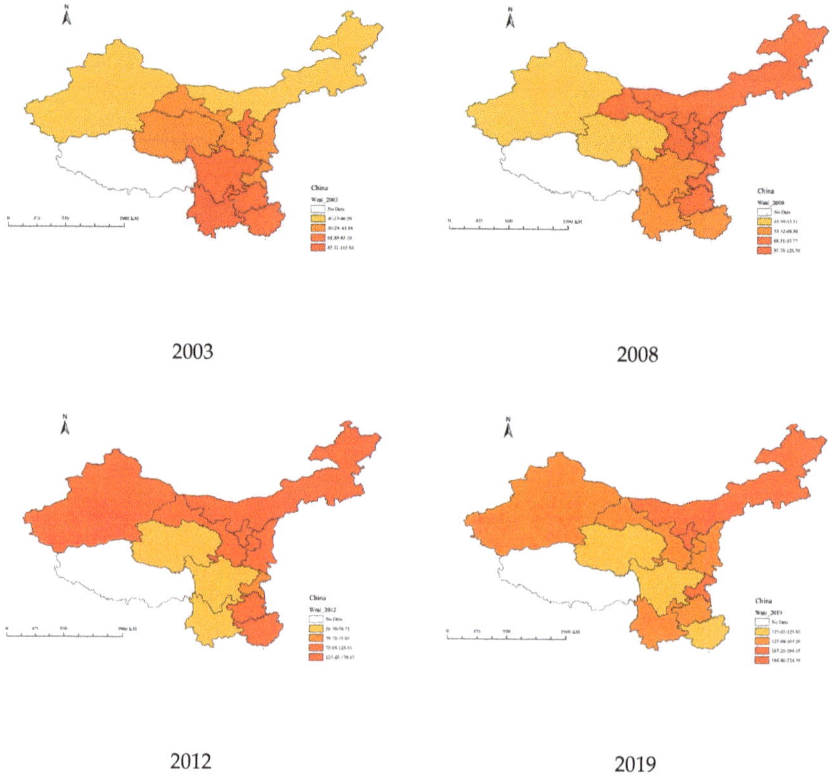

Figure 4. The level of market integration in western China.

In addition, after 2009, the market integration level in the northeast region significantly increased but sharply declined after 2016 in Figure 5. This may be because the promotion of the revitalization strategy of the northeast stimulated cooperation and exchange between regions to a certain extent. Still, due to the solidification of the industrial foundation and many difficulties in transformation, the industrial transformation may not be able to match further the integration process in the late stage of the market integration process. As a result, the level of market integration in the region showed a downward trend in the late sample period. The level of market integration in the western region changed more slowly before 2012, while it showed a significant upward trend after 2012 and a slight decline after 2016. This may be caused by the fact that most western regions belong to important ecological protection areas, which, to a certain extent, restricts the possibility of inter-regional economic collaboration, and guides these regions to invest more in environmental protection.

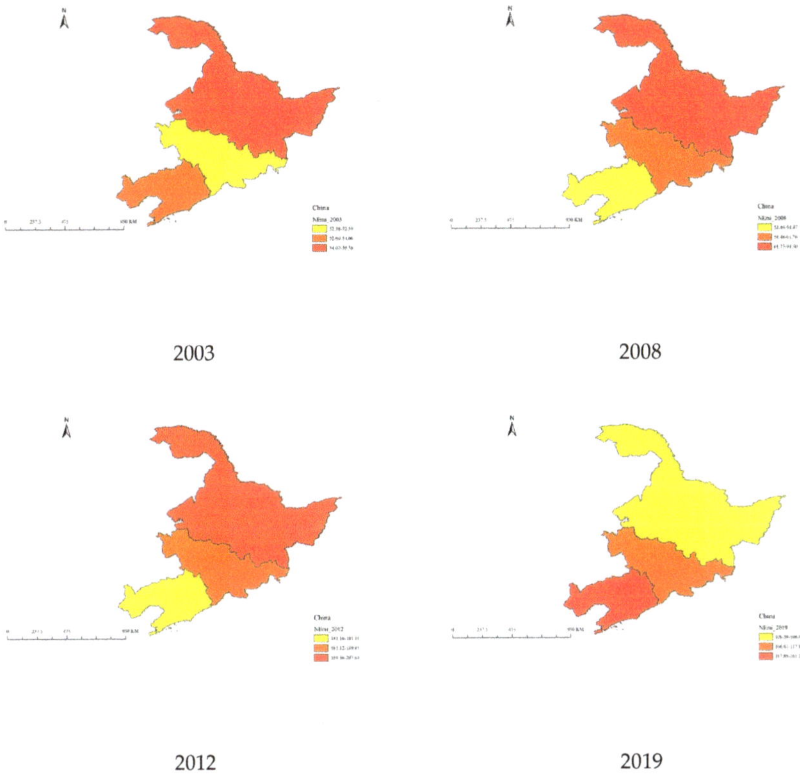

Figure 5. The level of market integration in Northeastern China.

4.1.2. Spatial and Temporal Trends in Carbon Emission

Figures 6–10 visualize the level of carbon emissions in each region, and it is evident from the data that the extent of the carbon emissions is closely linked to the area. It can be seen from the emissions that Hebei, Shanxi, Shandong, and Inner Mongolia are among the country's top regions in terms of carbon emission levels, which is also related to the pillar industries and development patterns of the areas in Figure 6. For example, in the results of the sub-region, heavy industries, such as coal mines, metal, and petroleum have relatively high carbon emissions.

Eastern China has always maintained the rising trend of carbon emission levels, keeping the second position in Figure 7. Obviously, as an economically developed region, the eastern region maintains a high level in terms of carbon emissions, both in terms of production activities and living agglomeration, but with the transformation and upgrading of the economic structure, more and more energy-consuming industries gradually move to the central and western regions, thus making the total amount of carbon emissions in the eastern region show a slight downward trend.

The carbon emission level in central China has shown a slight downward trend since 2012 in Figure 8, resulting from the fact that most provinces in central China are located in the Yangtze River economic belt, the ecological protection requirements of which may have a particular impact on the carbon emission level in central China.

It is noteworthy that western China's carbon emission level has always shown a significant upward trend in Figure 9, although its carbon emission level ranked last in 2013. However, by 2019, western China's overall carbon emission level had jumped to the top of the four areas. On the one hand, with the deepening of the western development strategy, the development of western China has shown an upward trend. On the other hand, along with the phenomenon of industrial transfer, the level and scale of industries undertaken by the part of the west in eastern and central China are also rising, which leads to a gradual increase in the carbon emission level of western China.

The northeast region has maintained a higher carbon emission position for a long time in Figure 10, which may be related to its regional characteristics; it carries more old industrial industries, and the transformation of economic and industrial development after 2012 may have had a particular impact on its carbon emission level. It is evident in the graph that the carbon emission level has fallen back since 2012.

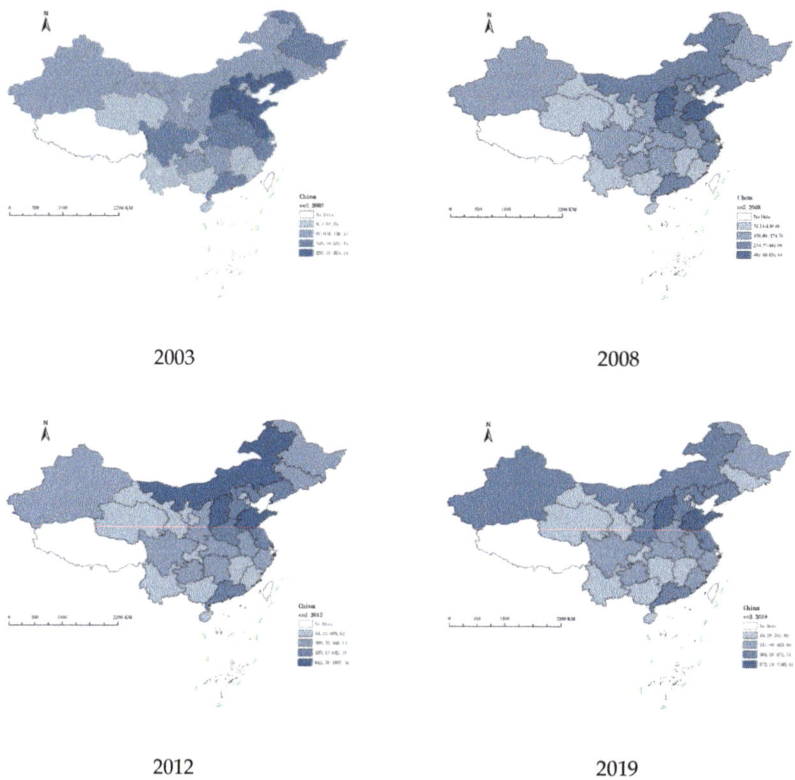

Figure 6. The level of carbon emissions in China.

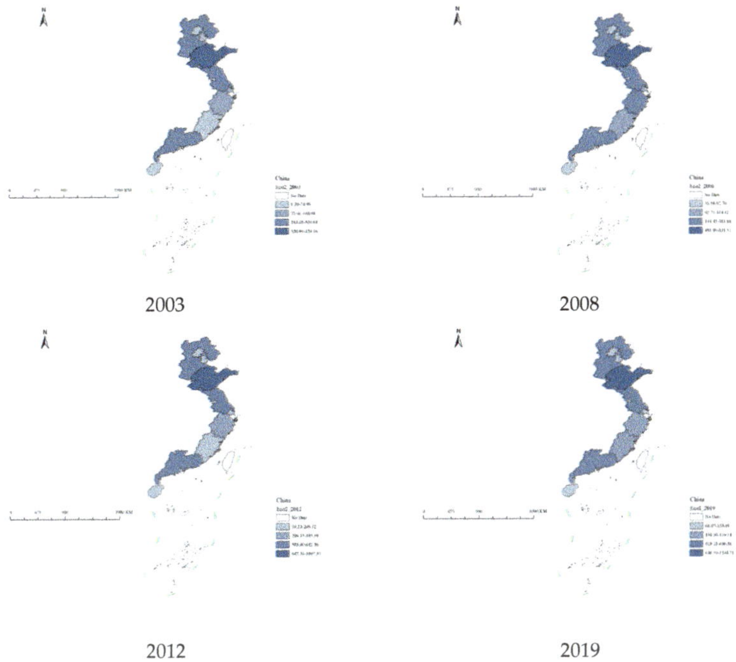

Figure 7. The trend of carbon emission levels in eastern China.

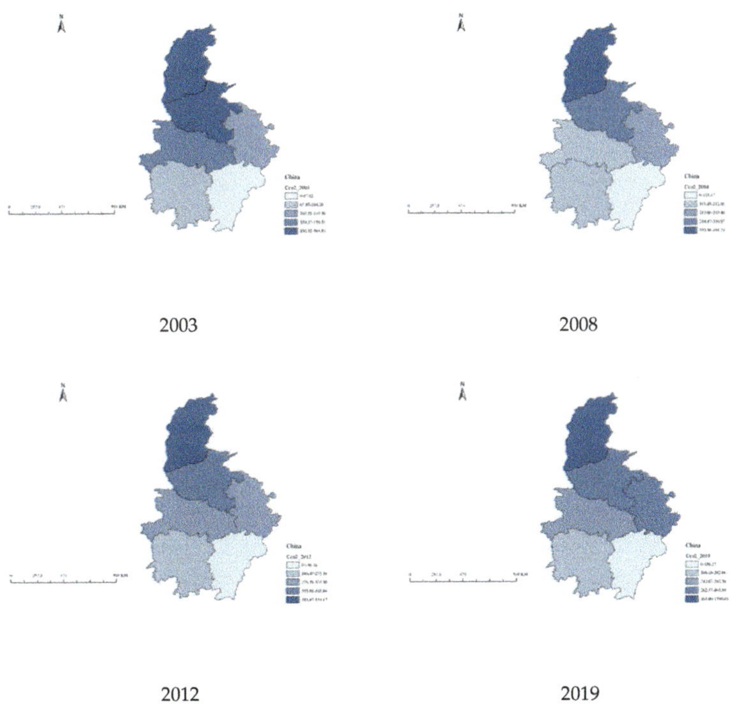

Figure 8. The trend of carbon emission levels in the central China.

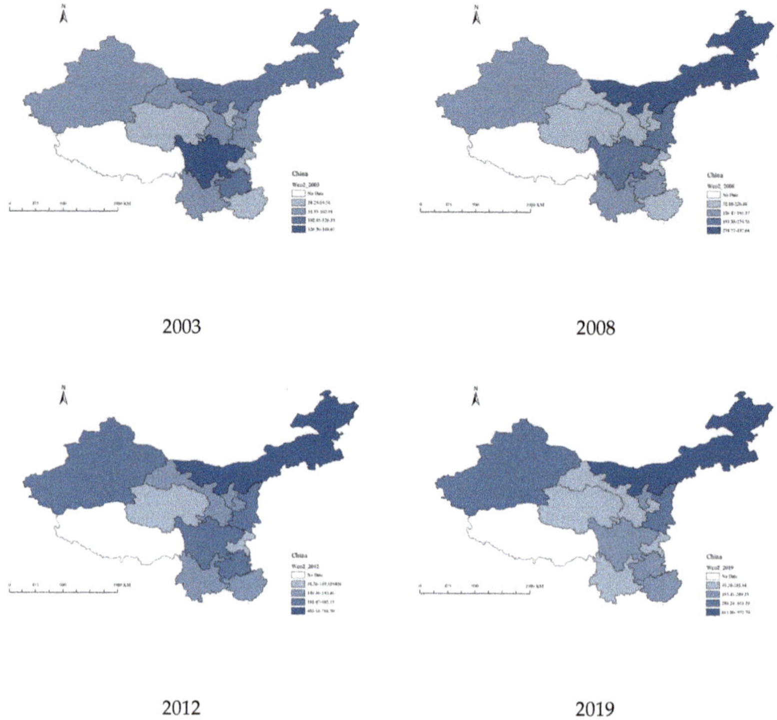

Figure 9. The trend of carbon emission levels in western China.

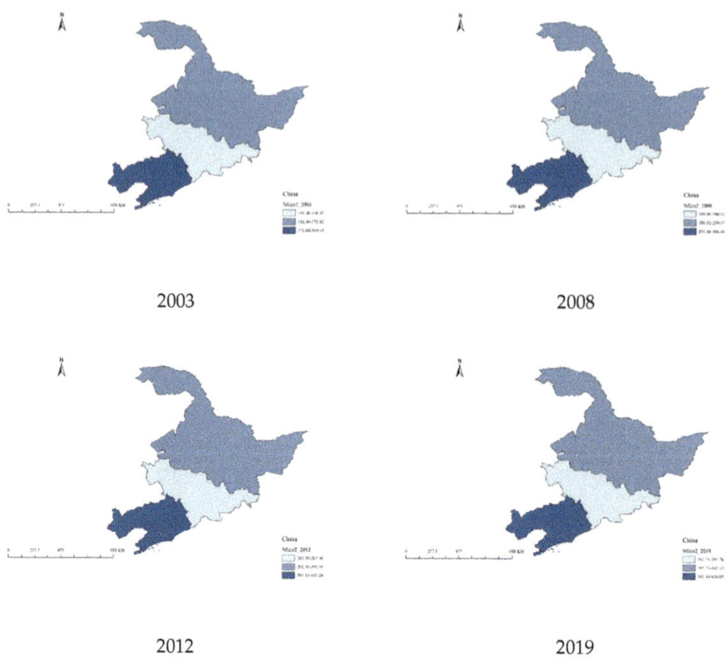

Figure 10. The trend of carbon emission levels in northeast China.

4.2. Analysis of Empirical Results

4.2.1. Regression Analysis Results

Table 1 indicates that under the baseline regression, market integration significantly increases carbon emissions without considering the impact of changes in industry structure. However, its impact is far less than that of government input (i.e., each 1% increase in the level of market integration will lead to a 0.0787% increase in carbon emissions), while every 1% increase in government input will lead to a 0.507% increase in carbon emissions. The increase in the level of market integration will help break down local barriers, accelerate the flow of various factors, and reduce transaction costs, thus promoting the scale effect to increase total production and consumption. Therefore, with other conditions unchanged, enterprises can produce more goods, and consumers' willingness and ability to pay are also enhanced to a certain extent. The increase in these production factors and the acceleration of commodity flow will increase carbon emissions [47].

Table 1. Results of the benchmark return.

	ALL TCE	EAST TCE	CENTRAL TCE	WEST TCE	NE TCE
MI	0.0787 **	−0.0262	0.437 ***	0.0111	0.0361
	(2.18)	(−0.50)	(4.04)	(0.25)	(0.87)
GOVI	0.507 ***	0.543 ***	0.289	0.605 ***	0.0150
	(4.78)	(3.23)	(1.15)	(5.56)	(0.10)
ZLSP	−0.0328	−0.141 **	0.151 *	−0.183 ***	0.0213
	(−0.88)	(−2.30)	(1.83)	(−4.46)	(0.43)
FDI	0.173	5.832 ***	−3.447 **	−0.962 *	3.315 ***
	(0.36)	(4.22)	(−2.41)	(−1.83)	(3.61)
GDP	−0.0615	−0.328 *	−0.0308	0.158	0.114
	(−0.47)	(−1.71)	(−0.10)	(1.05)	(0.61)
_cons	1.917 *	−10.22 ***	9.606 ***	3.900 ***	−4.761 *
	(1.66)	(−2.90)	(2.93)	(3.42)	(−2.00)
N	510	170	102	187	51
r2	0.765	0.725	0.732	0.860	0.850

Notes: t statistics in parentheses. * $p < 0.1$, ** $p < 0.05$, *** $p < 0.01$.

To further analyze whether there is regional heterogeneity in the impact of market integration on carbon emissions, regression analysis is carried out for eastern, central, western, and northeastern China without considering the factors of the industrial structure change. The results show that the impact of market integration on carbon emissions presents significant regional differences. The eastern, western, and northeastern regions have no significant impact, and only in the central area has market integration promoted the level of carbon emissions. The elasticity coefficient is 0.437; (i.e., for every 1% increase in market integration, carbon emissions will increase by 0.437%). As for other control variables, both government input and foreign investment levels in eastern China significantly contribute to the increase in carbon emissions, with elasticities of 0.543 and 5.832, respectively. In contrast, the technology and local economic development levels have a suppressive effect on carbon emissions, with elasticities of −0.141 and −0.328, respectively. Only western China negatively impacts carbon emissions in government input, with an elasticity coefficient of 0.605. At the same time, technology and foreign investment have a positive relationship, with elasticities of −0.183 and 0.962, respectively. Northeast China only shows a negative relationship with carbon emissions in foreign investment, with an elasticity coefficient of 3.315.

It suggests that the advancement of market integration in eastern China may not substantially impact carbon emissions. Meanwhile, it may be due to the higher level of economic development in eastern China, which has achieved inter-regional opening and integration earlier than other regions of the country, and the barriers between regions are relatively low. On the contrary, factors such as technology levels and government input may have a more significant impact on carbon emissions, while we consider that changes in industrial structure may have a relatively more significant impact on regional carbon emissions. Since 2005, the strategy for the rise of central China has led to a gradual increase in inter-regional cooperation. Therefore, in this context, the further opening of the market has accelerated the flow of factors and commodity transactions in central China, which has led the region to increase the amount of carbon emissions. At the same time, along with the requirements of national green development, central and western China have gradually tightened their investment requirements. As a result, it is not difficult to understand that the western and central regions have significantly suppressed the increase in carbon emissions in terms of foreign investment, but with the deepening of regional cooperation and further breaking of market barriers in central China, the spillover of technology is more likely to be concentrated in non-high-tech. The development of the technology level in this region is mainly applied to industrial production sectors such as manufacturing, thus showing that technological progress has significantly promoted the increase in carbon emissions.

4.2.2. Analysis of the Mechanism

To examine the moderating role played by industrial structure changes in the market integration process on carbon emissions, the interaction terms of market integration and industrial rationalization, and the interaction terms of market integration and industrial advancement, are incorporated into the model, respectively. The results are shown in Table 2.

Table 2. Analysis of moderator effects.

	(1) TCE	(2) TCE	(3) TCE	(4) TCE
MI	0.0798 **	0.0949 ***	0.0629 *	0.0633 *
	(2.21)	(2.66)	(1.79)	(1.79)
INDR	0.0916	0.160		
	(0.62)	(1.10)		
INDH			−0.250 ***	−0.250 ***
			(−5.17)	(−5.18)
GOVI	0.501 ***	0.459 ***	0.491 ***	0.492 ***
	(4.70)	(4.37)	(4.75)	(4.76)
ZLSP	−0.0301	−0.0160	−0.0674 *	−0.0666 *
	(−0.81)	(−0.44)	(−1.84)	(−1.81)
FDI	0.169	0.0717	−0.213	−0.216
	(0.35)	(0.15)	(−0.45)	(−0.46)
GDP	−0.0637	−0.115	−0.125	−0.127
	(−0.49)	(−0.90)	(−0.99)	(−1.00)
Intreact1		0.692 ***		
		(4.23)		
Intreact2				0.0148
				(0.26)
_cons	−0.0148	−0.118	−0.228 *	−0.230 *
	(−0.13)	(−1.00)	(−1.89)	(−1.90)
N	510	510	510	510
r2	0.765	0.774	0.778	0.778

Notes: t statistics in parentheses. * $p < 0.1$, ** $p < 0.05$, *** $p < 0.01$.

Columns (1) and (2) show the results of the effect of market integration on carbon emissions after considering industrial rationalization. In the main effects regression, the impact of market integration on carbon emissions is significantly negative, and the coefficient of the effect of market integration on carbon emissions is 0.0798, which means that every 1% increase in the level of market integration will lead to a 0.0798% increase in the level of carbon emissions. After adding the interaction term, the result of the interaction term is significantly negative; That is, industrial rationalization further enhances the impact of market integration on carbon emissions. This indicates that as the level of industrial rationalization among regions increases, it helps accelerate the optimization of the inter-regional division of labor and efficient collaboration, and can effectively improve the utilization of resources and energy. However, more production and consumption may increase carbon emissions and weaken the carbon reduction effect brought by improving resource and energy efficiency. Columns (3) and (4) show the results of the impact of market integration on carbon emissions after considering advanced industrialization. In the main effects regression, the effect of market integration on carbon emissions is also significantly negative; namely, every 1% increase in market integration will lead to a 0.0629% increase in carbon emissions. In addition, industrial upgrading has a significant positive effect on carbon emissions; every 1% increase in industrial upgrading will lead to a 0.25% decrease in carbon emissions.

It suggests that although industrial upgrading can play its unique advantage and role in promoting carbon emission reduction, it may play a limited role in market integration. With the promotion of market integration, different regions may have different paths and ways to achieve industrial upgrading. In promoting market integration, due to differences in levels or similar political demands, the strategic objectives and positioning of industrial restructuring may have a convergence effect. Therefore, industrial upgrading has not played a role in strengthening or weakening the impact of market integration on carbon emissions.

4.2.3. Intermediary Effect Analysis

We used industrial rationalization and industrial upgrade as mediating variables to explore whether market integration has a mediating effect on carbon emissions, and tested that market integration can additionally affect carbon emissions by influencing changes in industrial structure. The results of Table 3 indicate that using industrial upgrade as a mediating variable, it passes the Sobel Test (i.e., market integration can further affect the carbon emission level by influencing the change of industrial upgrading). The results show that the proportion coefficient of the intermediary effect of market integration on carbon emissions is 0.6654. Although the impact of market integration on carbon emissions is still negative after considering the level of an industrial upgrade as a mediating variable, it can be seen from the coefficient change that the coefficient of the effect of market integration on carbon emissions decreases from 0.397 to 0.268 with the intervention of industrial upgrade (i.e., every 1% increase in market integration level will lead to a 0.268% increase in carbon emissions). However, the carbon emission level will increase by 0.397% without industrial upgrade intervention. Interestingly, in the intermediary transmission process, we find that market integration has a negative effect on industrial upgrading, which may indicate that in promoting market integration and removing barriers between regions, the rapid transformation to industrial upgrading may not be achieved in the short term. The rapid development of local trade, industry, etcetera may be preferred, which will hinder or slow down the process of industrial structure upgrading. This phenomenon can be seen in the above regional heterogeneity analysis, and the process of industrial advancement is not the same between regions. Market integration cannot achieve a pull effect on the overall industrial progress in the short term. However, it is undeniable that market integration can still weaken carbon emissions by upgrading industrial structures. The rationalization of industrial structures failed to pass the intermediary effect test, indicating that regions may still fail to take industrial rationalization as the primary choice for regional industrial development in promoting market integration. Therefore, this invisible competition relationship may still exist among regions. This relationship cannot effectively promote the scientific and reasonable change of industrial structure, and there are risks of resource reuse, resource waste, and carbon emissions in the market integration process.

Table 3. Results of mediating effects of industrial upgrades.

	(1) TCE	(2) INDH	(3) TCE
MI	0.397 ***	−0.279 ***	0.268 ***
	(4.78)	(−3.50)	(3.560)
INDH			−0.462 ***
			(−10.94)
GOVI	0.39 ***	0.957 ***	0.832 ***
	(3.19)	(8.14)	(7.120)
ZLSP	−0.245 ***	−0.0896 *	−0.287 ***
	(−4.66)	(−1.77)	(−6.050)
FDI	−4.243 ***	2.429 ***	−3.121 ***
	(−7.56)	(4.50)	(−6.080)
GDP	1.023 ***	−1.128 ***	0.502 ***
	(10.09)	(−11.57)	(4.890)
_cons	5.585 ***	0.498	5.815 ***
	(4.48)	(0.420)	(5.20)
Sobel Test	0.129 ***		
	(3.331)		
Indirect effect	0.2682		
Direct effect	0.3972		

Notes: t statistics in parentheses. * $p < 0.1$, ** $p < 0.05$, *** $p < 0.01$.

4.2.4. Robustness Test

We tested the robustness of the benchmark regression by replacing variables, changing measures standards, extracting years, and other methods in Table 4. The results show that the impact of market integration on carbon emissions is always negative, but only slightly varies in the magnitude and coefficient of significance, indicating that the empirical results are reliable. Column (1) tests the model by the GLS method, and the results show that the effect of market integration on carbon emissions is significantly negative (i.e., every 1% increase in market integration will lead to a 0.0282% increase in the level of carbon emissions). Column (2) shows that after bringing the market integration lag into the model as the core explanatory variable, the effect of market integration on carbon emissions is still significantly negative (i.e., every 1% increase in market integration will lead to a 0.0282% increase in carbon emission level). Column (3) shows that after adjusting the data by excluding the data from 2003 and from 2019, the model regression results still indicate that the effect of market integration on carbon emissions is significantly negative (i.e., every 1% increase in market integration will lead to a 0.0738% increase in the level of carbon emissions). The results of the robustness test on the model indicate the stability of the regression analysis and the reliability of the results.

Table 4. Robustness test results.

	(1) TCE	(2) TCE	(3) TCE
MI	0.0282 ***	0.0818 **	0.0738 **
	(0.0095)	(0.0348)	(0.0369)
Control Variables	YES	YES	YES
Constant	1.9195 ***	2.4699 **	2.3137 *
	(0.3527)	(1.1846)	(1.2899)
N	510	450	450
r2		0.6822	0.7349

Notes: t statistics in parentheses. * $p < 0.1$, ** $p < 0.05$, *** $p < 0.01$.

5. Conclusions

This paper adopts a two-way fixed effects analysis based on Chinese provincial panel data from 2003 to 2019. It uses a mediation model to explore the relationship between market integration and carbon emissions, the carbon emission effects generated in promoting market integration between regions, and the shock effects caused by changes in industrial structure based on industrial rationalization and upgrade. The specific conclusions are as follows:

First, market integration significantly increases carbon emissions, however, there was significant regional heterogeneity. While the advancement of market integration in eastern China may not have a substantial impact on carbon emissions, the further opening of markets in central China accelerates the flow of factors and trade of commodities in central China, making this region demonstrate characteristics of market integration significantly increasing the number of carbon emissions.

Second, market integration will increase the scale of production through industrial upgrades, which will further increase carbon emissions; the results of the moderating effect show that market integration has a significant negative contribution on carbon emissions, and that industrial structure change can play a moderating role. Industrial rationalization can further enhance this negative effect. However, industrial upgrading does not have an effective regulatory result on this effect, but it significantly positively impacts carbon emissions. Therefore, it can be seen that along with improving industrial rationalization, it is conducive to improving the efficiency of production division and accelerating the circulation of commodity factors among regions. The increase in carbon emissions brought by this scale efficiency may be much more significant than the reduction effect brought by the improvement of resource and energy utilization efficiency, thus further enhancing the impact of market integration on carbon emission. As for industrial advancement, its effects on carbon emissions are more reflected in itself, but the role played by industrial upgrades in the process of market integration may be limited.

Third, industrial upgrading can reduce carbon emissions, but there are still some obstacles to market integration in the process of promoting industrial restructuring. We find that market integration has a negative effect on industrial upgrading, which may indicate that promoting market integration between regions and breaking down inter-regional barriers may not realize the rapid transformation to industrial upgrading in the short term. Market integration cannot promote the upgrading of the overall industry in a short time. However, it is undeniable that market integration can still weaken carbon emissions by upgrading the industrial structure. The rationalization of the industrial structure failed to pass the intermediary effect test, indicating that in promoting market integration, regions may still fail to take industrial rationalization as the primary choice for regional industrial development. Therefore, this invisible competition relationship may still exist among regions. This relationship cannot effectively promote the scientific and reasonable change of industrial structure, and there are risks of resource reuse, resource waste, and carbon emissions in the process of market integration.

Based on these findings, we propose several recommendations to the Chinese government to promote the development of a low-carbon economy and achieve China's carbon reduction targets.

First, the government should consider the two-way effect of low-carbon and market integration. With the expansion of market scale, the increase in production brought by the increase in the degree of market integration will undoubtedly lead to a short-term increase in carbon emissions. Therefore, it is necessary to take full advantage of the policy dividend of market integration to accelerate the rationalization of industrial transfer and locally advanced upgrading between areas to improve the various levels of technical exchanges, applications, and complementarities among regions.

Second, in the process of promoting market integration, each region should choose an industrial change path that is in line with the actual regional development, consolidate the various primary conditions for industrial development, achieve the valid promotion of increment in actual production and development, and effectively realize the recyclable mode of emission reduction. In particular, the central government should accelerate the implementation of diversified local assessment standards, optimize local officials' promotion options and ways as soon as possible, and ultimately break the rough inspection system of regional economic development only.

Finally, in promoting market integration, we should fully use spillover advantages, play the role of inter-regional transmission of technological innovation, and realize a good model of increasing output without increasing carbon in the market integration process. At the same time, central and western China should further strengthen the management of foreign direct investment, attach importance to the introduction and support of green industries, and form a sustainable industrial development pattern.

Author Contributions: Conceptualization, K.Z. and H.D.; methodology, K.Z.; software, K.L.; validation, K.Z. and S.Y.; formal analysis, K.Z. and Y.C.; investigation, K.Z.; resources, K.Z. and S.Y.; data curation, K.L.; writing—original draft preparation, K.Z.; writing—review and editing, K.Z.; visualization, K.Z.; supervision, K.Z.; project administration, H.D.; funding acquisition, H.D. All authors have read and agreed to the published version of the manuscript.

Funding: This research was funded by National Social Science Foundation later stage support project (2020) "Study on ecological civilization construction in River Basin" (20FGLB017).

Data Availability Statement: Not applicable.

Conflicts of Interest: The authors declare no conflict of interest.

References

1. Grossman, G.; Krueger, A.B. Environmental Impacts of a North American Free Trade Agreement. *C.E.P.R. Discuss. Pap.* **1991**, *3914*, 1–39. [CrossRef]
2. Antweiler, W.; Copeland, B.R.; Taylor, M.S. Is free trade good for the environment? *Am. Econ. Rev.* **2001**, *91*, 877–908. [CrossRef]
3. Cole, M.A.; Elliott, R.J.R. Determining the trade-environment composition effect: The role of capital, labor and environmental regulations. *J. Environ. Econ. Manag.* **2003**, *46*, 363–383. [CrossRef]
4. Managi, S.; Hibiki, A.; Tsurumi, T. Does trade openness improve environmental quality? *J. Environ. Econ. Manag.* **2009**, *58*, 346–363. [CrossRef]
5. Baek, J.; Cho, Y.; Koo, W.W. The environmental consequences of globalization: A country-specific time-series analysis. *Ecol. Econ.* **2009**, *68*, 2255–2264. [CrossRef]
6. Aller, C.; Ductor, L.; Herrerias, M.J. The world trade network and the environment. *Energy Econ.* **2015**, *52*, 55–68. [CrossRef]
7. Le, T.H.; Chang, Y.; Park, D. Trade openness and environmental quality: International evidence. *Energy Policy* **2016**, *92*, 45–55. [CrossRef]
8. Zhang, K.; Shao, S.; Fan, S. Market integration and environmental quality: Evidence from the Yangtze river delta region of China. *J. Environ. Manag.* **2020**, *261*, 110208. [CrossRef]
9. Sommer, K.; de Groot, H.L.; Klaassen, F. *The Effects of Market Integration on Pollution: An Analysis of EU Enlargements*; Discussion Paper; Tinbergen Institute: Amsterdam, The Netherlands; Rotterdam, The Netherlands, 2022.
10. Jiborn, M.; Kander, A.; Kulionis, V.; Nielsen, H.; Moran, D.D. Decoupling or delusion? Measuring emissions displacement in foreign trade. *Glob. Environ. Change* **2018**, *49*, 27–34. [CrossRef]
11. Shapiro, J.S.; Walker, R. Why Is Pollution from US Manufacturing Declining? The Roles of Environmental Regulation, Productivity, and Trade. *Am. Econ. Rev.* **2018**, *108*, 3814–3854. [CrossRef]
12. Van Biesebroeck, J. Exporting raises productivity in sub-Saharan African manufacturing firms. *J. Int. Econ.* **2005**, *67*, 373–391. [CrossRef]
13. Li, J.L.; Lin, B.Q. Does energy and CO_2 emissions performance of China benefit from regional integration? *Energy Policy* **2017**, *101*, 366–378. [CrossRef] [PubMed]
14. He, W.J.; Wang, B.; Danish; Wang, Z.H. Will regional economic integration influence carbon dioxide marginal abatement costs? Evidence from Chinese panel data. *Energy Econ.* **2018**, *74*, 263–274. [CrossRef] [PubMed]
15. Lin, B.Q.; Du, K.R. Energy and CO_2 emissions performance in China's regional economies: Do market-oriented reforms matter? *Energy Policy* **2015**, *78*, 113–124. [CrossRef]
16. Wang, W.; Zhang, Y.-J. Does China's carbon emissions trading scheme affect the market power of high-carbon enterprises? *Energy Econ.* **2022**, *108*, 105906. [CrossRef]
17. Chen, Z.; Liu, Y.; Zhang, Y.; Zhong, Z. Inter-regional economic spillover and carbon productivity embodied in trade: Empirical study from the Pan-Yangtze River Delta Region. *Environ. Sci. Pollut. Res.* **2021**, *28*, 7390–7403. [CrossRef]
18. Song, M.; Xu, H.; Shen, Z.; Pan, X. Energy market integration and renewable energy development: Evidence from the European Union countries. *J. Environ. Manag.* **2022**, *317*, 115464. [CrossRef]

19. Aalto, P. Energy Market Integration in East Asia: Deepening Understanding and Moving Forward. *J. Southeast Asian Econ.* **2015**, *32*, 181–183.
20. Wang, C.J.; Wang, F.; Zhang, X.L.; Yang, Y.; Su, Y.X.; Ye, Y.Y.; Zhang, H.G. Examining the driving factors of energy related carbon emissions using the extended STIRPAT model based on IPAT identity in Xinjiang. *Renew. Sustain. Energy Rev.* **2017**, *67*, 51–61. [CrossRef]
21. Wang, S.J.; Liu, X.P.; Zhou, C.S.; Hu, J.C.; Ou, J.P. Examining the impacts of socioeconomic factors, urban form, and transportation networks on CO_2 emissions in China's megacities. *Appl. Energy* **2017**, *185*, 189–200. [CrossRef]
22. Liao, B.; Li, L. How can urban agglomeration market integration promote urban green development: Evidence from China's Yangtze River Economic Belt. *Environ. Sci. Pollut. Res.* **2022**, *29*, 10649–10664. [CrossRef] [PubMed]
23. Jihong, Y.; Xiqiang, C. Whether regional integration cooperation leads to pollution transfer: Evidence from the enlargement of the Yangtze River Delta urban cluster. *China Popul. Resour. Environ.* **2019**, *29*, 118–129. [CrossRef]
24. Johansson, A.C.; Ljungwall, C. Spillover Effects Among the Greater China Stock Markets. *World Dev.* **2009**, *37*, 839–851. [CrossRef]
25. Grossman, G.M.; Helpman, E. Globalization and Growth. *Am. Econ. Rev.* **2015**, *105*, 100–104. [CrossRef]
26. Poncet, S. Measuring Chinese domestic and international integration. *China Econ. Rev.* **2003**, *14*, 1–21. [CrossRef]
27. Ke, S. Domestic Market Integration and Regional Economic Growth-China's Recent Experience from 1995–2011. *World Dev.* **2015**, *66*, 588–597. [CrossRef]
28. Krugman, P. Increasing returns and economic geography. *J. Political Econ.* **1991**, *99*, 483–499. [CrossRef]
29. Dong, B.M.; Gong, J.; Zhao, X. FDI and environmental regulation: Pollution haven or a race to the top? *J. Regul. Econ.* **2012**, *41*, 216–237. [CrossRef]
30. Yuping, D.; Helian, X. Foreign Direct Investment, Local Government Competition and Environmental Pollution:Empirical Analysis on Fiscal Decentralization. *China Popul. Resour. Environ.* **2013**, *23*, 155–163. [CrossRef]
31. Zhenxin, W.; Xiaojing, X.; Shuping, W. The Influence of Economic Development and Industrial Structure to Carbon Emission Based on China's Provincial Panel Data. *Chin. J. Manag. Sci.* **2012**, *20*, 161–166. [CrossRef]
32. Dong, B.; Ma, X.; Zhang, Z.; Zhang, H.; Chen, R.; Song, Y.; Shen, M.; Xiang, R. Carbon emissions, the industrial structure and economic growth: Evidence from heterogeneous industries in China. *Environ. Pollut.* **2020**, *262*, 114322. [CrossRef] [PubMed]
33. Chen, X.; Chen, Y.E.; Chang, C.P. The effects of environmental regulation and industrial structure on carbon dioxide emission: A non-linear investigation. *Environ. Sci. Pollut. Res.* **2019**, *26*, 30252–30267. [CrossRef] [PubMed]
34. Yi, M.; Liu, Y.; Sheng, M.S.; Wen, L. Effects of digital economy on carbon emission reduction: New evidence from China. *Energy Policy* **2022**, *171*, 113271. [CrossRef]
35. Young, A. The razor's edge: Distortions and incremental reform in the People's Republic of China. *Q. J. Econ.* **2000**, *115*, 1091–1135. [CrossRef]
36. Yu, Z.; Jiancai, P. Environmental Effects on Segmentation and In tegration of Regional Markets in China: From Perspective of Transboundary Pollution. *Financ. Econ.* **2013**, *57*, 101–110.
37. Shan, Y.; Huang, Q.; Guan, D.; Hubacek, K. China CO_2 emission accounts 2016–2017. *Sci. Data* **2020**, *7*, 54. [CrossRef]
38. Shan, Y.; Liu, J.; Liu, Z.; Xu, X.; Shao, S.; Wang, P.; Guan, D. New provincial CO_2 emission inventories in China based on apparent energy consumption data and updated emission factors. *Appl. Energy* **2016**, *184*, 742–750. [CrossRef]
39. Parsley, D.C.; Wei, S.J. Explaining the border effect: The role of exchange rate variability, shipping costs, and geography. *J. Int. Econ.* **2001**, *55*, 87–105. [CrossRef]
40. Chunhui, G.; Ruogu, Z.; Dianfan, Y. An Empirical Study on the Effects of Industrial Structure on Economic Growth and Fluctuations in China. *Econ. Res. J.* **2011**, *46*, 4–16+31.
41. Wei, L.; Hui, Z.; Zewei, H. An Examination of China's Industrial Structure Height and Industrialization Process and Regional Differences. *Econ. Perspect.* **2008**, *5*, 4–8.
42. Shioji, E. Public capital and economic growth: A convergence approach. *J. Econ. Growth* **2001**, *6*, 205–227. [CrossRef]
43. Wei, Z.Y. Examining the impact factors of energy-related CO_2 emissions using the STIRPAT model in Guangdong Province, China. *Appl. Energy* **2013**, *106*, 65–71. [CrossRef]
44. Sadorsky, P. Do urbanization and industrialization affect energy intensity in developing countries? *Energy Econ.* **2013**, *37*, 52–59. [CrossRef]
45. Xie, R.; Fang, J.; Liu, C. The effects of transportation infrastructure on urban carbon emissions. *Appl. Energy* **2017**, *196*, 199–207. [CrossRef]
46. Chitnis, M.; Sorrell, S.; Druckman, A.; Firth, S.K.; Jackson, T. Turning lights into flights: Estimating direct and indirect rebound effects for UK households. *Energy Policy* **2013**, *55*, 234–250. [CrossRef]
47. Liu, Z.; Zhang, H.; Zhang, Y.-J.; Duan, F.-E.; Wei, L.-Y. Research on the effects of market integration on carbon emissions. *Manag. Decis.* **2019**, *59*, 747–763. [CrossRef]

energies

MDPI

Article

How Does the Digital Economy Affect Carbon Emission Efficiency? Evidence from Energy Consumption and Industrial Value Chain

Kangni Lyu, Shuwang Yang, Kun Zheng * and Yao Zhang *

School of Economics and Management, China University of Geosciences, Wuhan 430074, China
* Correspondence: jgzk@cug.edu.cn (K.Z.); zhangyao@cug.edu.cn (Y.Z.)

Abstract: China is confronted with the dual constraints of economic transformation and carbon emission reduction. As the digital economy is a key force in promoting economic transformation and optimizing industrial structure, it is crucial to analyze the digital economy's impact on carbon emission reduction from the perspective of energy consumption and industrial value chain implications. We selected data from 251 prefecture-level cities and above in China from 2011 to 2019 as research samples, measured the development level of the digital economy using the entropy value method, and constructed relevant regression models based on two-way fixed effects, intermediary analysis, and moderation analysis. The research reveals that: (1) The digital economy has a significant contribution to carbon emission efficiency, and there are significant regional heterogeneity and city size differences; (2) The digital economy can improve carbon emission efficiency by reducing energy consumption. (3) From a value chain perspective, industrial structure rationalization weakens the carbon emission efficiency improvement effect of the digital economy to a certain extent, whereas industrial structure upgrading obviously enhances the carbon efficiency improvement effect of the digital economy. The above findings enrich the research in the field of digital economy and environmental governance, contribute to a more comprehensive understanding of the mechanisms by which the digital economy affects the carbon emission efficiency, as well as provide policy implications for enhancing the use of the digital economy in the regional energy consumption and industrial value chain.

Keywords: digital economy; carbon emission efficiency; industrial structure; energy consumption; industrial value chain; mediation model; moderation model

Citation: Lyu, K.; Yang, S.; Zheng, K.; Zhang, Y. How Does the Digital Economy Affect Carbon Emission Efficiency? Evidence from Energy Consumption and Industrial Value Chain. *Energies* **2023**, *16*, 761. https://doi.org/10.3390/en16020761

Academic Editor: Donato Morea

Received: 15 December 2022
Revised: 2 January 2023
Accepted: 4 January 2023
Published: 9 January 2023

1. Introduction

The effects of climate change on the sustainable growth of human society are significant. The Paris Agreement establishes the objective of managing the global temperature increase: accomplishing the goal of reducing global average temperature rise to no more than 2 °C and seeking to maintain it under 1.5 °C in order to safeguard the earth's ecological security [1]. To attain this long-term goal, nations must immediately peak their greenhouse gas emissions and contribute to the realization of carbon neutrality by the middle of this century. With China being the largest emitter of carbon dioxide and having the most comprehensive range of industries, China's effective promotion of low-carbon transition development is an obvious choice for deploying carbon peaks and carbon neutrality. A binding target of 13.5% reduction in energy consumption and 18% reduction in carbon emissions has been set for China under the 14th Five-Year Plan. In this situation, it is necessary to improve the efficiency of carbon emissions, achieving more economic growth with less energy and the same amount of carbon emissions, push for a complete low-carbon transformation in economic and social fields, and follow a green, low-carbon, high-quality development path to meet the "double carbon" target on time.

Currently, the digital economy is becoming increasingly influential in restructuring factor resources, changing economic structures, and transforming the competitive environment. According to the data released in *the Report on the Development of China's Digital Economy (2022)*, the digital economy in China achieved a new high in 2021, with its size reaching 45.5 trillion yuan with a nominal growth rate of 16.2%, 3.4% higher than the nominal growth rate of GDP, and accounting for 39.8% of GDP [2], strengthening its position in the national economy as well as playing an increasingly important supporting role.

Although the digital economy can provide long-term benefits to economic development, its possible environmental implications have also garnered the attention of numerous academics. In terms of theoretical and empirical studies, there is still a lack of direct discussion on the impact of the digital economy on the efficiency of carbon emissions, but relevant studies on the impact of the digital economy, especially on carbon emissions, provide references and inspirational implications for this paper. On the one hand, some scholars believe that the digital economy is an effective way to mitigate carbon emissions. They argue that the digital economy is providing a new impetus for intelligent management of the environment, with information technology at its core, and has a positive effect on environmental pollution control by functioning as informal environmental regulation [3,4]. Simultaneously, the extrusion effect of the digital economy can effectively promote the transformation and upgrading of the regional industrial structure, further restrain the development of high energy-consuming and high-polluting industries, and thus accelerate the improvement of environmental quality [5,6]. On the other hand, an opposite viewpoint on the impact of the digital economy on carbon emissions has gained significant attention. Proponents of this view believe that the development of the digital economy does not reduce energy consumption, but rather increases it, and that the energy growth effect of the digital economy may have outweighed the energy reduction effect [7]. Moreover, the expansion of the digital economy increases the size of the economy, which in turn increases energy consumption and carbon emissions [8]. These contradicting findings have piqued the interest of academics in researching the impact of the digital economy on carbon emissions.

Research on carbon emission efficiency is mainly focused on the measurement of carbon emission efficiency among different regions and industries, and the analysis of the influencing factors of carbon emission efficiency [9,10]. Research on digital economy and carbon emission mainly focuses on the digital economy's impact on carbon emission [11,12], the link between the digital economy and carbon emission performance [13], and the impact of internet development on carbon emission efficiency [14]. However, few scholars have argued for a possible direct impact relationship between carbon emission efficiency and the digital economy, and there is a lack of further evidence to support the relationship in terms of value chain and energy consumption. This paper therefore seeks to bring the variety of digital economy development into the research framework of the impact factor theory for carbon emission efficiency in order to determine whether the digital economy influences carbon emission efficiency, if such an effect exists. What is the mechanism of industrial value chain and energy consumption in the process if this effect does exist? To answer the above questions, this paper combines the distinctive characteristics of the digital economy and constructs a theoretical analysis framework from the perspective of carbon emission efficiency. Based on this framework, the digital economy and carbon emission efficiency levels of 251 prefecture-level and above cities in China were measured from 2011 to 2019, and the impact of the digital economy on urban carbon emission efficiency and its mechanism of action were empirically examined. The findings indicated that the digital economy greatly improved carbon emission efficiency in the region, with the reduction of energy consumption and the improvement of industrial value chains being among the most significant mechanisms of effect.

As a crucial component of high-quality development, the influence of the digital economy on economic growth and ecological environment is gaining increasing attention. As a typical developing country, examining the contribution of China's digital economy to enhancing carbon emission efficiency can provide developing countries with theoretical support for enhancing the carbon emission reduction capacity of the digital economy and achieving "economic growth—environmental protection" win-win development. The original study's potential marginal contributions included the following three aspects: (1) This paper empirically tested whether the digital economy had a positive impact on carbon efficiency using data from prefecture-level municipalities, providing new empirical evidence for research related to the digital economy and environmental quality, especially in the area of carbon emissions. It also offered potential policy references for green development in China; (2) A comprehensive evaluation index system at the municipal level was constructed, therefore enhancing the measurement approach. To comprehensively reflect the development of the digital economy in Chinese cities, comprehensive digital economy indicators were constructed with the internet as the core, and the characteristics of digital economy development and the influence relationship between the digital economy and carbon emission efficiency were discussed in greater detail. (3) We explored the intrinsic mechanism of the digital economy affecting carbon emission efficiency, and used energy consumption as a mediating variable to analyze the transmission path, effect size and heterogeneous differences of the digital economy on carbon emission efficiency improvement, and clarify the policy focus points for promoting the low-carbon development of the digital economy.

The remainder of this paper is structured as follows. Section 2 presents the relevant research hypotheses through literature analysis. Section 3 provides an overview of the research data and methodology. Sections 4 and 5 explain and analyze the empirical results. Section 6 elaborates the conclusions and policy recommendations of this study.

2. Research Hypotheses

The impact routes of the digital economy on carbon emission efficiency may be analyzed from two perspectives: the direct impact road from the standpoint of digital economy development, and the indirect impact path from the standpoints of energy consumption and the industrial value chain. The influence mechanism of the digital economy on carbon emission efficiency is depicted in Figure 1.

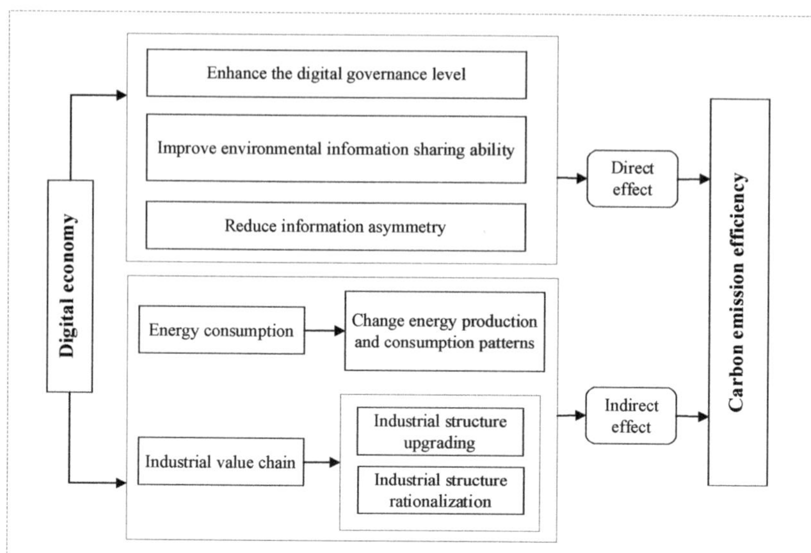

Figure 1. Mechanism analysis between digital economy and carbon emission efficiency.

2.1. The Direct Impact of the Digital Economy on the Efficiency of Carbon Emissions

Numerous scholars have studied the energy and environmental effects of the digital economy's development and found that the rapid development of the digital economy exemplified by the internet not only results in rapid economic growth, but also contributes to significant improvements in environmental performance [15,16]. It has been determined that the rapid expansion of the digital economy, as demonstrated by the internet, brings not only quick economic growth but also major improvements in environmental performance. The rise of the digital economy has a substantial impact on carbon emissions, the most evident signal of change within the framework of climate change. On the one hand, the development of the digital economy drives up the level of digital technology, the application of digital technology in environmental protection has changed the traditional environmental monitoring model and method by combining various sensors and computer technology to create a comprehensive network information collection system, realizing the integration of data collection and transmission and management, reducing the cost of monitoring technology, and enhancing the monitoring capability of real-time assessment of environmental conditions [17]. The efficient sharing of environmental information facilitates effective resource deployment, compensates for the deficiencies of traditional regulatory tools in a targeted manner, provides data support to enhance environmental regulation and enforcement, and thereby improves pollution management [18]. Moreover, the development of digital technology offers new options and avenues for business information disclosure, thereby mitigating the negative effects of information asymmetry [19]. In addition, it strengthens the competition mechanism of elimination of winners and losers in the market environment, forcing enterprises with high pollution and high emissions to invest more in research and development to achieve an efficient use of resources and low carbon and sustainable development of the city [20]. On the other hand, the development of the internet has brought about changes in connectivity and communication, accelerated the speed of information transfer, enriched access to information, provided more opportunities for knowledge sharing, use and re-creation, enabled traditional industries to take advantage of the penetration and derivation of digital technology for industrial upgrading, promoted the process of technical catch-up and economic convergence, as well as the development of intelligent and environmentally friendly industries [21], and reduced energy consumption and pollutant emissions [22]. In addition, the efficiency of carbon emission is improved. In summary, we propose Hypothesis 1.

Hypothesis 1 (H1). *The digital economy positively affects carbon emission efficiency.*

2.2. Indirect Impact of the Digital Economy on Carbon Efficiency

2.2.1. Digital Economy and Carbon Efficiency: The Energy Consumption Perspective

Energy consumption is a key driver of carbon emissions, which include the consumption of natural resources such as coal, oil and natural gas [23]. In the context of the expansion of the digital economy, an increasing number of studies have proven the function of the use of digital devices and processes that might increase energy efficiency in many industries [24–26]. Specifically, in promoting the integration of traditional energy companies with digital enterprises, the use of digital technologies has significantly improved the operational efficiency of oil and gas companies. Additionally, the latest information technology has been utilized to integrate energy and digital technologies in order to build a new energy ecosystem, change the way energy is produced and consumed, optimize the energy mix, accelerate the energy transition, and improve carbon emission performance [27,28]. The digital economy accelerates urban processes and brings about the development of public transportation and renewable energy [29], which helps to capitalize on the economies of scale of public infrastructure and prevent environmental damage [30]. Simultaneously, the extensive use of big data analysis can effectively promote the construction of the global energy internet, which can effectively improve the efficiency of energy resource allocation, enable the development and consumption of clean energy to reach scale, and gradually replace fossil fuel energy, which is conducive to reducing carbon dioxide emissions and can improve carbon emission efficiency [12,31].

Conversely, it has been proposed that rapid urban expansion and development increases intensive urban economic activities caused by housing, transportation and recreation [32], which increases energy demand and leads to more carbon emissions [33], which reduces the regional carbon efficiency, and the energy consumption associated with the creation of digital infrastructure itself may negate any possible energy savings. It is argued that the digital economy based on communication technologies has energy-intensive qualities, and a huge quantity of infrastructure construction will consume more energy resources in the early stages of digital economy development [7]. In addition, data creation, process, storage, and movement depend on resources such as water, electricity, and metals, and as the scale of use of digital services and products continues to expand, the environmental pollution caused by e-waste during use and carbon emissions also increase [34]. Collard et al. and Longo et al. also believe that the usage of ICT has resulted in an increase in electricity consumption and that communication technologies have not significantly improved the environment [35,36]. Therefore, we argue that the digital economy can affect carbon efficiency through influencing energy consumption, and in this paper, we propose the following mediation hypothesis.

Hypothesis 2 (H2). *Energy consumption plays a mediating role between the digital economy and carbon efficiency.*

2.2.2. Digital Economy and Carbon Efficiency: Industrial Value Chain Perspective

In the context of the rapid development of information technology, the internet, with its characteristics of openness, collaboration and sharing, has gradually become the most important production application tool, and its integration with traditional production factors and resources has promoted industrial upgrading [37]. Gereffi et al. suggest that industrial upgrading can be seen as a process of climbing up the value chain or between value chains for firms and the whole industry in the region [38]. The productivity dividend brought by the deep integration of new generation communication technology and advanced manufacturing technology can significantly break the innovation bottleneck of each link in the industrial chain, thus breaking the "low-end locking" trap of the industrial value chain and making the industrial structure develop from low-level to high-level forms with inter-industrial upgrading [39], and the degree of change from low to high value-added industries can directly reflect the quality and level of development of the industrial value chain.

The inter-industrial upgrade will, to a certain extent, diminish the good impact of the digital economy's development on reducing carbon emission efficiency. The rapid emergence and evolution of digital technology has created a new opportunity for the industrial structure to transform from a factor-driven to an innovation-driven mode. This can help boost sectoral productivity and improve the industrial value chain [40]. Moreover, digital network platforms can promote resource sharing among industries and fields via scale and competition effects, optimize traditional industrial production and sales methods, strengthen the market competition mechanism, eliminate backward production capacity, and force backward and low-end industries to upgrade [41]. Existing scholars have argued that inter-industrial upgrading might successfully cut carbon emissions via a variety of techniques [42,43]. In the process of developing industrial structure in a green direction, the fossil energy-based energy structure will be significantly enhanced, especially for energy-intensive and carbon-emitting industrial sectors, and digital technology will reduce the demand for energy and materials, which can effectively improve energy efficiency and resource allocation efficiency, thereby reducing carbon emissions. Therefore, we believe that inter-industry upgrading is an important element in industrial value chain upgrading, as the impact of the digital economy on carbon emission efficiency will be significantly influenced by industrial structure value-chain upgrading. Furthermore, we divide the inter-industry upgrading into two dimensions, industrial structure advanced and industrial structure rationalization. The industrial structure upgrade process involves increasing the number of high-value-added industries. This process is carried out to improve the overall structure of the facility. The second is industrial structure rationalization; the higher the degree of inter-industry coordination, the higher the degree of industrial structure rationalization.

From the above, we propose the following hypotheses.

Hypothesis 3a (H3a). *Industrial structure upgrading plays a moderating role between digital economy and carbon efficiency.*

Hypothesis 3b (H3b). *Industrial structure rationalization plays a moderating role between digital economy and carbon efficiency.*

3. Model

3.1. Method

To test the above research hypotheses, a two-way fixed-effects model is first constructed for the direct transmission mechanism.

$$eff_{it} = \alpha_0 + \alpha_1 digital_{it} + \alpha_2 X_{it} + \mu_i + \delta_t + \varepsilon_{it} \quad (1)$$

In Equation (1), eff_{it} represents the carbon emission efficiency of city i at time t, $digital_{it}$ is the digital economy development index of city i in period t, X_{it} is a vector that represents the remaining control variables, μ_i is the individual fixed effect, δ_t refers to the time-fixed effect, and ε_{it} denotes the random error term.

Besides the direct effect embodied in Equation (1), this study also explored the possibility that the consumption of energy can be a factor mediating the digital economy and carbon emissions. Referring to the stepwise method proposed by Baron and Kenny (1986) for testing mediating effects [44]: the coefficient α_1 significance of the model (1) of digital economic development index (*digital*) on carbon emission efficiency (eff) passed the test, so we constructed linear regression equations for *digital* on the mediating variable energy consumption (*energy*), as well as regression equations for *digital* and the mediating variable *energy* on eff. The mediation effect will be judged by the significance of regression coefficients such as β_1, γ_1 and γ_2. The following is the specific form of the regression model:

$$energy_{it} = \beta_0 + \beta_1 digital_{it} + \beta_2 X_{it} + \mu_i + \delta_t + \varepsilon_{it} \quad (2)$$

$$eff_{it} = \gamma_0 + \gamma_1 digital_{it} + \gamma_2 energy_{it} + \gamma_3 X_{it} + \mu_i + \delta_t + \varepsilon_{it} \quad (3)$$

Further, this section adds the interaction term of industrial structure upgrading (*insu*) and industrial structure rationalization (*inso*) with digital economy development index (*digital*) to test the role of industrial structure moderation between digital economy and carbon emission efficiency, the significance of the regression coefficients such as η_3 and η_7 will be used to determine whether the moderating effect exists.

$$eff_{it} = \eta_0 + \eta_1 digital_{it} + \eta_2 insu_{it} + \eta_3 digital_{it} \times insu_{it} + \eta_4 X_{it} + \mu_i + \delta_t + \varepsilon_{it} \quad (4)$$

$$eff_{it} = \eta_0 + \eta_5 digital_{it} + \eta_6 inso_{it} + \eta_7 digital_{it} \times inso_{it} + \eta_8 X_{it} + \mu_i + \delta_t + \varepsilon_{it} \quad (5)$$

3.2. Variables

3.2.1. Dependent Variable

The explanatory variable studied in this paper is carbon emission efficiency (eff). This research is based on the super-efficient SBM model proposed by tone [45], which incorporates labor input, capital stock, and energy consumption as input indicators, GDP as desired output and carbon emissions as non-desired output, as stated in Table 1. (1) Labor input is indicated by the number of employees in each prefecture-level city at the end of the year. (2) The estimation of capital stock is mostly calculated using the perpetual inventory method at constant prices, and this part draws on Zhang's approach [46], which adopts a discount rate of 9.6% to calculate the capital stock at the end of each year from 2011 to 2019, using the year 2000 as the base period, the calculation formula is $K_{it} = K_{it-1}(1 - \delta_{it}) + l_{it}$, where K_{it} is the capital stock of region i in year t. l_{it} is the fixed asset investment of region i in year t. δ_{it} is the depreciation rate. (3) The direct energy consumption of the city mainly includes natural gas and liquefied petroleum gas, whereas the indirect energy consumption mainly includes electricity consumption, which will be converted into standard coal by referring to the *General rules for calculation of the comprehensive energy consumption GBT2589-2020* because units are not uniform. (4) The estimation of carbon dioxide emissions is based on the approach of shan, according to *the Intergovernmental Panel on Climate Change (IPCC)* guidelines on the allocation of greenhouse gas emissions, carbon emissions are calculated for 17 fossil fuel combustion and cement production-related process emissions for 47 socioeconomic sectors [47]. Figure 2 shows the spatial distribution of carbon emission efficiency indicators, and most cities show an increasing trend of carbon emission efficiency. Due to the limitation of space, this paper only provides the calculation results of two years.

Table 1. Evaluation system of carbon emission efficiency.

Input/Output	Indicators	Definition	Units
Input	Labor force	Number of employees in the unit at the end of year	10,000 people
	Capital stock	Total fixed assets at the end of year	10,000 yuan
	Energy consumption	Total energy consumption of natural gas, liquefied petroleum gas and electricity at the end of year	10,000 tons of coal
Desirable output	Economic output	Gross domestic product (GDP) at the end of year	10,000 yuan
Undesirable output	Carbon emission	Carbon emission at the end of year	10,000 tons

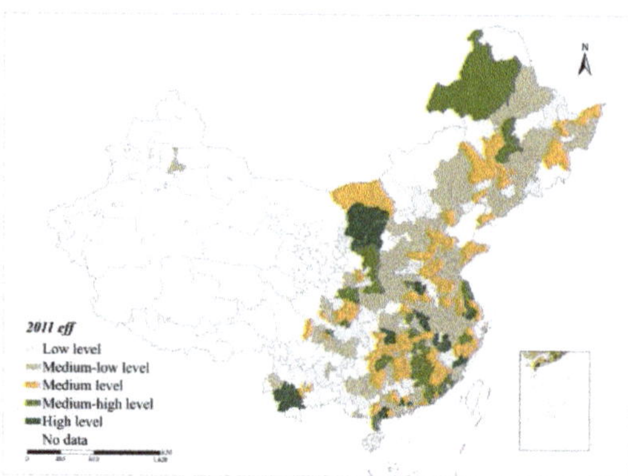

(a)

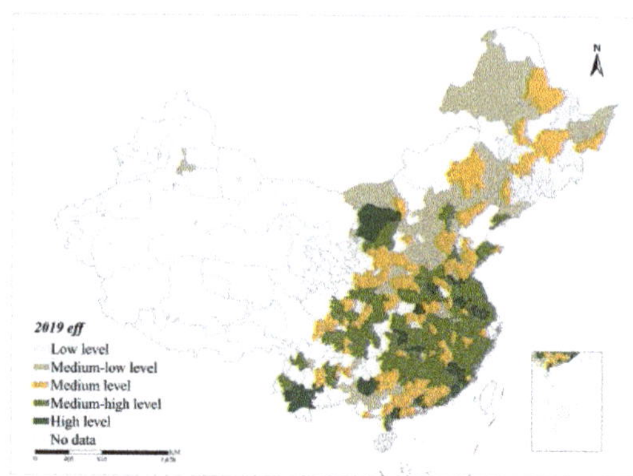

(b)

Figure 2. China's carbon emission efficiency. (**a**) Spatial distribution in 2011; (**b**) Spatial distribution in 2019.

3.2.2. Independent Variable

The digital economy index (*digital*) is the key explanatory variable for this article. Currently, there is a paucity of relevant research regarding the precise measurement of the digital economy, and academics have not yet developed a recognized evaluation system. Based on the study findings on the definition of digital economy, the design of an index system, and measurement methodologies, this work utilizes the availability of city-level data and the methodology of Zhao et al. to examine the economic and financial characteristics of the digital economy [48], measuring the development level of digital economy from internet development and digital finance. Considering the postal express business has increased fast in recent years along with the rapid development of e-commerce, the promotion of internet popularity on the scale of local postal express is relatively stronger than other factors [49]. In addition, the expansion of the digital economy has presented technology inventors with increasingly specialized business model difficulties, which in part encourages technology innovation [50]. In this article, internet penetration rate, internet-related practitioners, internet-related output, cell phone penetration rate, postal service output, and technological innovation capability are considered as indicators of internet development level. For digital finance development indicators, the Digital Financial Inclusion Index of China is used, which is compiled by *the Institute of Digital Finance Peking University* and *Ant Financial Group Holdings Limited*, comprehensively measuring three aspects: breadth of digital finance coverage, depth of use, and degree of digitalization [51]. The specific description is shown in Table 2. As an objective weighting method, the entropy method has a stronger objectivity, so this paper processes the data of the above indicators through the entropy method to obtain the digital economy index (*digital*). Figure 3 shows the trend of China's digital economy development level by cities. In general, digital economic development is more advanced in 2019 than it was in 2011.

Table 2. Evaluation system of the digital economy development index.

Primary Indicators	Secondary Indicators	Tertiary Indicators	Indicator Description	Unit	Indicator Properties
Digital economy development index	Internet development level	Internet penetration rate	Number of Internet access users per 100 people	household	+
		Internet-related practitioner	Computer services and software industry employees accounted for the proportion of urban unit employees	%	+
		Internet-related output	Total telecom services per capita	Yuan	+
		Cell phone penetration rate	Number of cell phone subscribers per 100 people	household	+
	Digital technology application	Postal operations output	Total postal services per capita	Yuan	+
		Technology innovation capability	Number of digital economy-related invention patent applications in the current year	Pieces	+
	Digital finance development level	Digital financial inclusion	Digital financial inclusion index of China	–	+

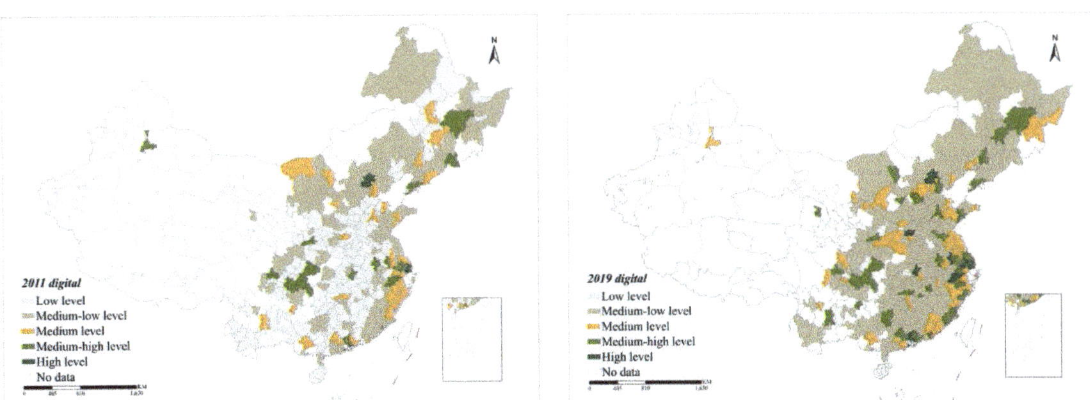

(a) (b)

Figure 3. China's digital economy. (**a**) Spatial distribution in 2011; (**b**) Spatial distribution in 2019.

3.2.3. Control Variables

To mitigate omitted variable bias as much as possible, the article further controls for a series of variables that affect the efficiency of urban carbon emissions. (1) Economic growth ($pgdp$), as measured by GDP per capital [50,52]; (2) Population density (pop), the ratio of total population to administrative area is chosen to represent population density [53,54]; (3) Environmental regulation (er), this paper collates all *Report on the Work of the Government* in prefecture-level cities from 2011–2019 by hand, sub-phrase the texts, count the frequency of environment-related words (Environment-related terms specifically include: pollution, emission reduction, prevention, ecological protection, low carbon, PM2.5, pm2.5, haze, emissions, emissions, air, blue sky defense war, pm10, PM10, green, environmental protection, particulate matter, monitoring, energy saving, dust, noise, tailpipe, emissions, environmental protection, forest coverage, soot, atmosphere, sulfur dioxide, SOD, ozone, sewage, SO_2, binding indicators, wastewater, recycling, water conservation, nitrogen oxides, energy, clean, unit GDP, chemical oxygen demand, energy consumption, ecological construction, green water and green mountains, low carbon, pollution control, waste gas, carbon dioxide, energy saving, ecology) and their proportion to the total number of words in the report, so as to characterize the environmental regulation [55]; (4) Foreign direct investment (fdi), as measured by the ratio of the annual actual foreign direct investment as a percentage of GDP [56,57]; (5) Financial development ($finan$), as calculated by the ratio of loan balances in financial institutions to regional GDP at the end of the year [58]; (6) Urban transportation network construction ($trans$), as measured by the road area per capita [59].

3.3. *Data Sources and Descriptive Statistics*

This research examines Chinese prefecture-level and higher cities between 2011 and 2019. Due to the challenges of incomplete data or poor data quality in some cities, the panel data of 251 cities are finally retained, and a small amount of missing data are compensated by linear interpolation. China City Statistical Yearbook, China Energy Statistical Yearbook, China Macro Economy Database, prefecture-level Municipal Statistical Bulletin, prefecture-level Report on the Work of the Government, website of Institute of Digital Finance Peking University, Carbon Emission Accounts and Datasets (CEADs), CSMAR database are the sources for the data used in this study. In addition, this paper uses the annual average price of RMB to USD exchange rate from the National Bureau of Statistics to adjust the total of foreign direct investment; the standard coal conversion is based on the general rules for calculation of the comprehensive energy consumption GBT2589-2020. In order to reduce the dispersion of the data, this paper logarithmically processes certain indicators. The results of descriptive statistics for the major variables in this work are presented in Table 3.

Table 3. Descriptive statistics of the variables (before logarithm).

Variables	Symbol	Obs	Mean	Std. Dev.	Min	Max
Carbon emission efficiency	*eff*	2259	0.45	0.19	0.17	1.38
Digital economy index	*digital*	2259	0.05	0.05	0.01	0.71
Economic growth	*pgdp*	2259	53,292.50	34,022.71	9773.00	467,749.00
Population density	*pop*	2259	3719.28	2536.95	179.00	15,055.00
Environmental regulation	*er*	2259	0.01	0.01	0.00	0.15
Foreign direct investment	*fdi*	2259	0.02	0.02	−0.03	0.20
Financial development	*finan*	2259	0.96	0.55	0.12	7.45
Urban transportation network	*trans*	2259	5.14	6.35	0.21	73.04

4. Results

4.1. *Baseline Regression Results*

Table 4 displays the results of the linear regression estimation of the digital economy affecting urban carbon emissions' efficiency. Models 1 and 2 show the results of fixed-effects model tests with and without control variables. Despite the absence of control factors, the digital economy can still help to improve carbon emission efficiency at a significant level of 1%. This is consistent with the conclusion of Hypothesis 1. Furthermore, there is a substantial positive correlation between the level of economic growth and carbon emission efficiency in Model 2, indicating that regional carbon emission efficiency has been effectively increased as a result of urban economic growth and economically developed regions with advanced production technology. This is probably because economic growth in China is gradually shifting from extensive to low-carbon model [10]. In contrast, the level of financial development and urban transportation network has a negative correlation with urban carbon emission efficiency. This may be because the financial development and the construction of the urban transportation network significantly accelerated the degree of urban development and expansion, which aggravated the total amount of carbon emissions in the region, thereby decreasing the efficiency of carbon emissions in a certain period of time [60,61].

Table 4. Baseline regression results and instrumental variable test results.

	Model 1 *eff*	**Model 2** *eff*	**Model 3** *digital*	**Model 4** *eff*
digital	0.7549 ***	0.5905 ***		3.2639 ***
	(7.9560)	(6.2796)		(3.4517)
distance × *mean_digital*			−0.0004 ***	
			(−7.4312)	
ln*pgdp*		0.1375 ***		0.0667
		(8.8389)		(1.3197)
ln*popud*		0.0028		0.0027
		(0.4188)		(0.4462)
er		−0.3187		0.3380
		(−0.9814)		(0.9002)
fdi		−0.1314		0.1350
		(−0.6896)		(0.6190)
finan		−0.0258 **		−0.0419
		(−2.5714)		(−1.6251)
ln*trans*		−0.0491 ***		−0.0132
		(−4.8218)		(−0.7097)
Constant	0.3960 ***	−0.9827 ***	0.0330 ***	−1.7502 ***
	(64.0975)	(−5.6587)	(18.6782)	(−3.6478)
KleibergenPaap rk LM statistic				27.57 [0.000]
KleibergenPaap rk Wald F statistic				26.74 {16.38}
Observations	2259	2259	2259	2259
Year	YES	YES	YES	YES
City	YES	YES	YES	YES
Adjust R^2	0.131	0.181	0.361	0.685
F	66.58	50.89	170.3	76.86

Notes: ***, **, and * indicate statistical significance at the 1%, 5%, and 10% levels. The figures in () are t statistics and in [] is P value of the corresponding test statistics. The critical value at the level of 10% critical values of Kleibergen-Paap rk Wald F test is within {} [62].

The findings of the baseline regression indicate a substantial positive correlation between the digital economy and carbon emission efficiency, and the development of regional digital economy contributes to the improvement of local carbon emission efficiency. The digital economy indicator system constructed in this paper may have measurement errors due to the availability of data, resulting in correlations between digital economy development indicators and unobservable factors affecting carbon emission efficiency. Besides, the reverse causality may exist between digital economy development and carbon emission efficiency.

This paper attempts an instrumental variable approach to mitigate the endogeneity problem. We use the spherical distance between each city and Hangzhou (The research methodology drawn from this paper selects the geographic feature of spherical distance from cities at all levels and above to Hangzhou as an instrumental variable. This instrumental variable is correlated with the degree of digital economy development in the region. The growth of digital finance exemplified by Alipay started in Hangzhou; thus, Hangzhou is the leading city in terms of digital economy development, and it is reasonable to predict that the closer a city is geographically to Hangzhou, the greater the level of digital economy development) as an instrumental variable (*distance*), and interact *distance* with the mean value of the digital economy development index (*mean_digital*) at the national level in the corresponding year as a new instrumental variable with time-varying effects [63]. Model 3 of Table 4 demonstrates that the estimated coefficient of the instrumental variable is −0.0004, which is statistically negative at the 1% significantly level. It implies that the more distant from the digital economy development center, the lower the level of the digital economy development, which is in line with expectations. After considering the endogeneity of the variables, the results of model (4) indicate that the digital economy still has a significant contribution to the efficiency of carbon emissions, which further supports the conclusion obtained from the benchmark regression, indicating that the improvement of the development level of the digital economy contributes to the improvement of carbon emission efficiency.

4.2. Robustness Tests

4.2.1. Dynamic Panel Regression

Different models have been selected to analyze and test the impact of the digital economy on carbon emission efficiency. One of the biggest issues in the estimation process of the model is the treatment of the endogeneity problem, as this endogeneity is caused by the system itself, which is identical to the dynamic panel data in this respect. This paper further uses the dynamic panel regression to test the robustness of the benchmark regression. The System Generalized Method of Moments (SYS-GMM) estimation is commonly utilized in dynamic panel data estimations to address endogeneity issues, and the SYS-GMM is compared to the difference Generalized Method of Moments (difference-GMM) by introducing level equations to reduce estimation errors. In order to evaluate the model, this study employs a two-stage SYS-GMM estimation approach; the estimation results are presented in Table 5. As can be seen from Model 1, AR (1) test rejects the null hypothesis at the 1% significance level, and AR (2) test cannot reject the null hypothesis, indicating that the model does not have higher-order serial correlation. The p-value of the Hansen test is 0.2080, which satisfies the over-identification test, indicating that the instrumental variables selected in this paper are reasonable and valid. The results from the SYS-GMM method demonstrate that the coefficients of the digital economy on carbon emission efficiency are significantly positive at the 1% level, which is consistent with the results of the baseline regression, supporting the robustness of the baseline regression.

Table 5. Robustness test results.

	Model 1 *eff*	Model 2 *eff*	Model 3 *eff*	Model 4 *eff*
L.eff	0.8280 ***			
	(21.2600)			
Score	0.3076 **	0.5905 ***	0.6276 ***	
	(2.0572)	(6.2796)	(5.5234)	
L.Score				0.9311 ***
				(8.8773)
Control variables	YES	YES	YES	YES
Constant	−0.5737	−0.9827 ***	−0.9797 ***	−1.2033 ***
	(−1.5285)	(−5.6587)	(−5.5661)	(−6.7456)
Observations	2008	2259	2223	2008
Year	YES	YES	YES	YES

Table 5. *Cont.*

	Model 1 *eff*	Model 2 *eff*	Model 3 *eff*	Model 4 *eff*
City	YES	YES	YES	YES
Hansen-p	0.2080			
AR (1)-p	0.0000			
AR (2)-p	0.3542			
F		50.89	47.50	55.69
Adjust R^2		0.1808	0.1689	0.2044

Notes: ***, **, and * indicate statistical significance at the 1%, 5%, and 10% levels. The figures in () are t statistics or z statistics.

4.2.2. Controlling Provincial Fixed Effect

Considering the possible changes in the macro-systemic environment caused by the widespread expansion of the digital economy, this section mitigates the possible changes of the digital economy development by introducing province-fixed effects and interaction effects between provinces and years. The estimation results of Model 2 in Table 5 show that the digital economy still plays a positive role in enhancing the carbon emission efficiency after considering the systematic changes of macro factors.

4.2.3. Excluding Municipalities Directly under the Central Government

Since Beijing, Tianjin, Shanghai, and Chongqing are under the direct jurisdiction of the central government, the administrative status is relatively special compared to other prefecture-level cities. In order to avoid the influence of administrative variables on the findings of the baseline regression, this section excludes the four municipalities from the full sample and then performs the regression test again. The estimation result in Model 3 of Table 5 shows that the regression coefficient is 0.6276, which is significantly positive at the 1% level, proving the robustness of the baseline regression results.

4.2.4. Replacing the Core Explanatory Variable

In consideration of the time required for the development of the digital economy to influence low-carbon development in the region by building infrastructure and restructuring industries, as well as to further mitigate the possible reverse causality, this paper treats the digital economy variables with a one-period lag. As shown by Model 4 in Table 5, after the lagged one-period treatment, the digital economy still contributes significantly to the carbon emission efficiency in the region at 1% level, which supports the results of the baseline regression.

4.3. *Heterogeneity Analysis*

Due to disparities in resource endowments and phases of development, there are obvious heterogeneous characteristics in the regional distribution of both digital economy development levels and carbon emission efficiency levels. This study examines regional differences in the impact of the digital economy on carbon emission efficiency at the city level from two perspectives: sub-regional and city-level, in light of the potential spatial heterogeneity of the impact of digital economy development on urban carbon emission efficiency. The regional classification is separated into four regions based on the regional location of each city in the province: northeastern, eastern, central and western regions. For the classification of city levels, the sample of central cities in this paper mainly includes municipalities directly under the central government, sub-provincial cities and provincial capitals, and other prefecture-level cities as peripheral cities. Before regression testing, descriptive statistics are performed on the disparities in digital economy development and carbon emission efficiency between regions and city levels. According to the descriptive statistics in Table 6, the eastern region is significantly ahead of other regions in terms of the degree of digital economy development, and the central cities have a "first mover advantage" over the peripheral cities; There is also some variation in the mean values of carbon emission efficiency between regions. The preceding conclusion lays the groundwork for testing the geographical heterogeneity of the digital economy and its impact on regional carbon emission efficiency.

Table 6. Regional digital economic development level and carbon emission efficiency.

	N	Mean	Std. Dev	Min	Max
			digital		
Northeastern	297	0.0407	0.0168	0.0124	0.108
Eastern	738	0.0727	0.0838	0.0117	0.714
Central	675	0.0398	0.0267	0.00913	0.348
Western	549	0.0410	0.0267	0.00936	0.229
Central Cities	288	0.122	0.117	0.0241	0.714
Peripheral Cities	1971	0.0406	0.0234	0.00913	0.272
			eff		
Northeastern	297	0.354	0.112	0.172	1.073
Eastern	738	0.489	0.182	0.234	1.230
Central	675	0.452	0.172	0.217	1.115
Western	549	0.466	0.220	0.176	1.375
Central Cities	288	0.434	0.189	0.176	1.202
Peripheral Cities	1971	0.458	0.186	0.172	1.375

Regression analysis of regional heterogeneity is performed in Figure 4. The regression results of line 1 to line 4 show that in the northeastern, eastern and central regions, the development of the digital economy plays a significant role in improving the carbon emission efficiency, especially in the eastern and central regions; the regression results are significantly positive at the 1% level. This suggests that the eastern region took initiatives in developing digital economy and has more obvious advantages in digital infrastructure and digital industry development, allowing them to play a greater role in digital empowerment with a variety of benefits, which is more important for carbon emission efficiency improvement. At the same time, the eastern region has a radiation-driven effect on the development of digital economy for the central region, benefiting from the digital technology spillover from the eastern region, the development pattern of the digital economy in the central region is further optimized, thus significantly contributing to the carbon efficiency of the central region. The northeastern region belongs to the traditional old industrial base area, along with the rapid development of the digital economy, the stimulating effect on the local traditional industrial sector may be more obvious, especially in promoting the digitalization of industrial industries. The northeastern region has been able to pay more attention to the use of low-carbon technologies in the process of industrial transformation and upgrading, which has greatly improved the efficiency of local carbon emissions. The digital economy development variables for the western region do not pass the significance test, most likely because the western region is still in the primary stage of digital economy development and the network infrastructure construction is still at a lagging level due to factors such as geographical location and factor accumulation. Lower resource utilization efficiency may be a significant factor as to why the digital dividend in the western area is not completely used.

The last two lines in Figure 4 indicate that the digital economy in central cities has a significant influence on improving carbon emission efficiency, whereas the development of the digital economy in peripheral cities has a significant inhibitory effect on regional carbon emission efficiency. This may be due to the fact that central cities have obvious advantages in the development process, exerting their siphon effect to gather various factors and forming basically perfect digital economy infrastructure, whereas peripheral cities are relatively backward in digital economy development and are still at the developing stage of digital economy. Furthermore, the construction of digital economy infrastructure in cities generates more resource consumption, which reduces the efficacy of the infrastructure.

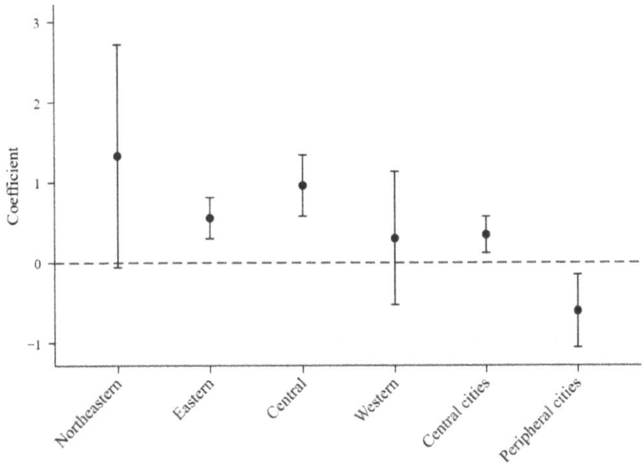

Figure 4. Regression results based on heterogeneity of sub-regional and city-level.

5. Discussion

Previous studies indicate a positive correlation between the digital economy and carbon emission efficiency, but the mechanisms by which the digital economy influences carbon emission efficiency still need to be further investigated. This section analyzes the transmission mechanism in greater detail to determine which factors can influence the digital economy and the carbon emission efficiency of the region.

5.1. Digital Economy, Energy Consumption and Carbon Emission Efficiency

In the previous chapters, we discussed possible mechanisms and pathways for the digital economy to influence carbon efficiency from the perspective of energy consumption. To verify this mechanism of action, we use energy consumption (Urban energy consumption mainly includes natural gas, liquefied petroleum gas and urban electricity. In this paper, the main urban energy consumption is converted into standard coal and then summed up to obtain urban energy consumption) as a mediating variable to test whether the digital economy has a further effect on carbon emission efficiency by influencing energy consumption. In Table 7, the results of Model 2 indicate that the coefficient of the digital economy is negative at the 1% level. This suggests that the development of the digital economy has a negative effect on energy consumption. It illustrates how the growth of the digital economy helps to utilize new energy sources and to enhance the efficiency of energy use to alleviate the problem of excessive energy consumption. The result of Model 3 shows that the coefficient of energy consumption is also significantly negative at the 1% level, whereas the coefficient of the digital economy is notably positive, which indicates that it has a favorable influence on carbon emission efficiency through optimizing the energy structure, confirmed by Hypothesis 2. This may be because the application of the digital economy in the energy sector accelerates the process of energy transition and improves the efficiency of energy production and utilization, which in turn reduces unnecessary energy consumption and improves the regional carbon emission efficiency.

Table 7. Intermediary effect regression results.

	Model 1 *eff*	Model 2 ln*energy*	Model 3 *eff*
digital	0.591 ***	−2.391 ***	0.326 ***
	(0.0940)	(0.340)	(0.0873)
ln*erengy*			−0.110 ***
			(0.00568)
Control variables	YES	YES	YES
Constant	−0.983 ***	1.722 ***	−0.792 ***
	(0.174)	(0.628)	(0.160)
Observations	2259	2259	2259
Adjust R^2	0.277	0.706	0.392
Year	YES	YES	YES
City	YES	YES	YES

Notes: ***, **, and * indicate statistical significance at the 1%, 5%, and 10% levels. The figures in () are t statistics.

5.2. Digital Economy, Industrial Value Chain and Carbon Efficiency

Although inter-industry upgrading promotes industrial value chain upgrading and further plays an important role in economic growth, it is equally important for improving carbon emission efficiency and promoting green development in China [64].

The transformation of industrial structure includes industrial structure upgrading (*isu*) (The upgrading of the industrial structure indicates the process of industrial structure's evolution and growth from a low level to a high level in accordance with the general rule of economic development.The research drawn from this part constructs the AIS index to calculate the advanced industrial structure by the cosine of the angle) and industrial structure rationalization (*iso*) (Industrial structure rationalization refers to the process of industrial restructuring and coordination. This paper draws on the practice of using the Theil index role to measure the degree of industrial structure rationalization), which are used as the moderating variables in the regression, respectively. The regression results are shown in Table 8. The result of model 2 indicates that the interaction term between industrial structure upgrading and digital economy is significantly positive at the 1% level, and industrial structure upgrading significantly enhances the influence of digital economy to promote carbon emission efficiency, which indicates that when industrial structure upgrading is at a high level, new industries with low energy consumption, low emission and high efficiency develop vigorously. Meanwhile, as the digital economy has grown, so has the need for digital management among local businesses, thereby creating good conditions for the region to use the digital economy to promote carbon efficiency. Thus, Hypothesis 3a is validated. The outcome of model 4 demonstrates, however, that the rationalization of industrial structure has a considerable weakening inhibitory impact in the process of promoting carbon emission efficiency by the digital economy, which confirms Hypothesis 3b. This indicates that the issue of adapting the development of the digital economy to the local industrial base and industrial structure is neglected in the process of promoting regional economic development, and the importance of the development of the digital economy is overemphasized, with more hotspot-oriented policy and other adjustments in the process of regional development. In fact, the rationalization of regional industries requires that the development of the digital economy must follow the objective laws of local economic and social development in order to achieve a reasonable allocation of production factors, which in turn can promote the coordinated development of various industries.

Table 8. Moderating effect regression results.

	Model 1 *eff*	Model 2 *eff*	Model 3 *eff*	Model 4 *eff*
digital	0.595 ***	−1.025 ***	0.591 ***	−0.465 ***
	(0.0951)	(0.255)	(0.0941)	(0.173)
insu	0.00968	0.0370		
	(0.0276)	(0.0276)		
digital × *insu*		1.746 ***		
		(0.255)		
inso			−0.0147	−0.0559 **
			(0.0245)	(0.0249)
digital × *inso*				−4.316 ***
				(0.596)
Control	YES	YES	YES	YES
Constant	−1.039 ***	−1.420 ***	−0.969 ***	−1.230 ***
	(0.236)	(0.240)	(0.175)	(0.177)
Observations	2259	2259	2259	2259
Adjust R^2	0.277	0.294	0.277	0.296
Year	YES	YES	YES	YES
City	YES	YES	YES	YES

Notes: ***, **, and * indicate statistical significance at the 1%, 5%, and 10% levels. The figures in () are t statistics.

Further, we analyze the relationship between the digital economy and carbon emission efficiency at the level of industrial structure upgrading and the level of industrial structure rationalization above or below the median. The results of Figure 5a indicate that the development of the digital economy has a significant effect on carbon emission efficiency in both high and low industrial structure upgrading. On the contrary, Figure 5b shows that the growth of the digital economy has a negative effect on carbon emission efficiency in both high and low industrial structure rationalization. This effect is particularly pronounced when the industrial structure rationalization at a high level.

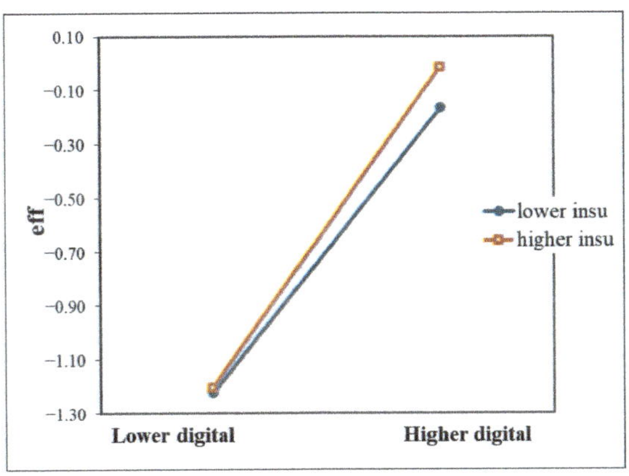

(a)

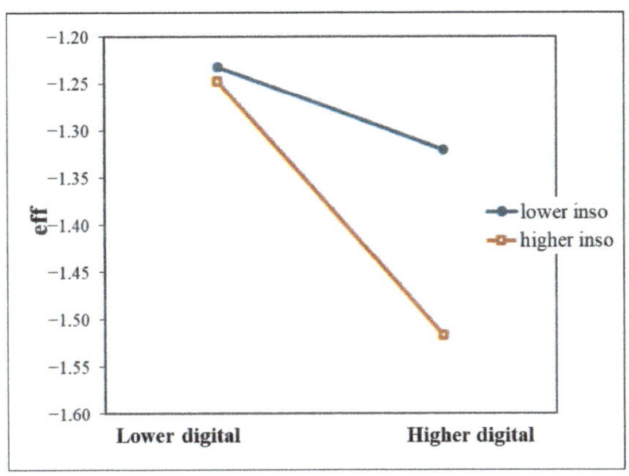

(b)

Figure 5. Moderating effect of industrial transformation. (**a**) Industrial structure upgrading; (**b**) Industrial structure rationalization.

6. Conclusions and Policy Implication

Based on the panel data of Chinese prefecture-level cities from 2011–2019, the carbon emission reduction mechanism and influence on the digital economy are empirically tested in several dimensions based on the construction of the digital economy development level index. The key findings are as follows: First, the digital economy significantly improves carbon emission efficiency, and the conclusions are still valid when endogeneity and a series of robustness variables are taken into account; Second, the effect of digital economy on carbon emission efficiency is regionally heterogeneous, with greater promotion effects in the eastern and central regions. The digital economy in central cities also has a significant effect on carbon emission efficiency, whereas peripheral cities on the contrary have a significant inhibitory effect; Third, the mechanism analysis shows that the digital economy can help improve the efficiency of urban carbon emissions by improving energy consumption as a pathway; Fourth, the industrial value chain has a moderating effect on the impact of digital economy on carbon emission efficiency, among which, industrial structure upgrading can significantly enhance the impact of digital economy on carbon emission efficiency enhancement, although it has a significant weakening and inhibiting effect in the process of digital economy promoting carbon emission efficiency enhancement. The main contribution of this paper is to provide more theoretical and empirical support for the influence of the digital economy on carbon emission efficiency. Nonetheless, there is potential for development in this paper, mostly owing to the availability of data. The assessment index system for the growth of the digital economy in cities is insufficiently thorough, so the index system presented in this study may also be inadequately extensive. Future enhancements of the established indicator system are contingent upon technological feasibility and data availability.

Based on the preceding facts, we conclude the following policy implications.

First, the digital economy is progressively becoming a significant driver of economic growth, and the findings of this paper imply that the expansion of the digital economy is also favorable to the accomplishment of the carbon peak carbon neutrality aim. Therefore, local governments should accelerate the growth of the digital economy and maximize the dividend impact of the digital economy on reducing carbon emissions efficiently. By constructing a high-speed, green and low-carbon, secure and controllable, intelligent and comprehensive digital information infrastructure, the government should accelerate the application of digital economy in social life, especially in the environmental field, how to guide the transformation and upgrading of traditional industries, and rely on digital technologies such as 5G, big data, and artificial intelligence to promote industrial innovation and pollution emission reduction, and foster the emergence of new technologies, industries, and business models related to low-carbon fields. Governments should leverage their scale impact and technology effect to transform the digital economy into a sustainable force that promotes carbon emission efficiency.

Second, the application of the internet in the energy industry should be enhanced and its integration with energy production and consumption should be encouraged. In the context of developing the digital economy, the government should guide the transformation and upgrading of high energy-consuming industries, especially by putting the digital economy technology represented by the Internet into the development and transformation of traditional manufacturing industries, encouraging the intelligent upgrading of energy production, transportation, consumption and other aspects, realizing the upgrading and optimization of industrial structure, promoting the deep integration of the Internet and the real economy, and pushing the further transformation of the manufacturing industry from traditional manufacturing to intelligent manufacturing. Promoting the low-carbon transformation of the energy sector can make full use of the urbanization process, apply the digital economy to the process, realize the adjustment and optimization of the energy structure, and promote the level of green ecological environment in cities.

Finally, digitalization can be used to promote regional green and coordinated development and reduce regional disparities. The digital economy gradually integrates local economic activities into regional production networks, leading to changes in regional production and industrial organization, and has become a significant vehicle for promoting urban and economic transformation. Although digitalization consolidates the advantages of digital economy development in eastern and central regions as well as in central cities, it also gradually promotes the transfer of digital economy technology inputs and applications to less developed regions, so that less developed regions can also receive technological dividends, achieve effective growth in economic efficiency in the region, and gradually reduce the overall gap with developed regions. At the same time, we should strengthen inter-regional cooperation, create an integrated digital economy intelligent service platform, establish a large digital economy service repository, realize data and technology resource sharing, use digital technology to enhance urbanization and digital governance in each region, and promote the development of the low-carbon economy.

Author Contributions: Conceptualization, K.L. and S.Y.; methodology, K.L. and Y.Z.; software, K.L.; validation, K.L. and S.Y.; formal analysis, K.L. and K.Z.; investigation, K.L.; resources, K.L. and S.Y.; data curation, K.Z.; writing-original draft preparation, K.L.; writing—review and editing, K.L. and Y.Z.; visualization, K.L.; supervision, K.L.; project administration, S.Y.; funding acquisition, S.Y. All authors have read and agreed to the published version of the manuscript.

Funding: This research was funded by the National Office for Philosophy and Social Sciences, grant number: 20FGLB038.

Data Availability Statement: Not applicable.

Conflicts of Interest: The authors declare no conflict of interest.

References

1. United Nations Framework Convention on Climate Change (UNFCCC). The Paris Agreement. Available online: Extension: //bfdogplmndidlpjfhoijckpakkdjkkil/pdf/viewer.html?file=https%3A%2F%2Funfccc.int%2Fsites%2Fdefault%2Ffiles%2Fenglish_paris_agreement.pdf (accessed on 11 November 2022).
2. China Academy of Information and Communications Technology (CAICT). Report on the Development of China's Digital Economy. 2022. Available online: http://www.caict.ac.cn/english/research/whitepapers/202208/t20220819_407677.html (accessed on 17 November 2022).
3. Usman, A.; Ozturk, I.; Hassan, A.; Zafar, S.M.; Ullah, S. The Effect of ICT on Energy Consumption and Economic Growth in South Asian Economies: An Empirical Analysis. *Telemat. Inform.* **2021**, *58*, 101537. [CrossRef]
4. Li, L.; Zheng, Y.; Zheng, S.; Ke, H. The New Smart City Programme: Evaluating the Effect of the Internet of Energy on Air Quality in China. *Sci. Total Environ.* **2020**, *714*, 136380. [CrossRef]
5. Lin, R.; Xie, Z.; Hao, Y.; Wang, J. Improving High-Tech Enterprise Innovation in Big Data Environment: A Combinative View of Internal and External Governance. *Int. J. Inf. Manag.* **2020**, *50*, 575–585. [CrossRef]
6. Zhu, J.; Xie, P.; Xuan, P.; Zou, J.; Yu, P. Renewable Energy Consumption Technology Under Energy Internet Environment. In Proceedings of the 2017 IEEE Conference on Energy Internet and Energy System Integration (ei2), Beijing, China, 26–28 November 2017; IEEE: New York, NY, USA, 2017.
7. Lange, S.; Pohl, J.; Santarius, T. Digitalization and Energy Consumption. Does ICT Reduce Energy Demand? *Ecol. Econ.* **2020**, *176*, 106760. [CrossRef]
8. Lee, C.-C.; Yuan, Z.; Wang, Q. How Does Information and Communication Technology Affect Energy Security? International Evidence. *Energy Econ.* **2022**, *109*, 105969. [CrossRef]
9. Zeng, L.; Lu, H.; Liu, Y.; Zhou, Y.; Hu, H. Analysis of Regional Differences and Influencing Factors on China's Carbon Emission Efficiency in 2005-2015. *Energies* **2019**, *12*, 3081. [CrossRef]
10. Sun, W.; Huang, C. How Does Urbanization Affect Carbon Emission Efficiency? Evidence from China. *J. Clean. Prod.* **2020**, *272*, 122828. [CrossRef]
11. Zhang, C.; Liu, C. The Impact of ICT Industry on COtextsubscript2 Emissions: A Regional Analysis in China. *Renew. Sustain. Energy Rev.* **2015**, *44*, 12–19. [CrossRef]
12. Yi, M.; Liu, Y.; Sheng, M.S.; Wen, L. Effects of Digital Economy on Carbon Emission Reduction: New Evidence from China. *Energy Policy* **2022**, *171*, 113271. [CrossRef]
13. Zhang, W.; Liu, X.; Wang, D.; Zhou, J. Digital Economy and Carbon Emission Performance: Evidence at China's City Level. *Energy Policy* **2022**, *165*, 112927. [CrossRef]
14. Wang, J.; Dong, K.; Sha, Y.; Yan, C. Envisaging the Carbon Emissions Efficiency of Digitalization: The Case of the Internet Economy for China. *Technol. Forecast. Soc. Chang.* **2022**, *184*, 121965. [CrossRef]
15. Lin, B.; Zhou, Y. Does the Internet Development Affect Energy and Carbon Emission Performance? *Sustain. Prod. Consump.* **2021**, *28*, 1–10. [CrossRef]
16. Wu, L.; Zhang, Z. Impact and Threshold Effect of Internet Technology Upgrade on Forestry Green Total Factor Productivity: Evidence from China. *J. Clean. Prod.* **2020**, *271*, 122657. [CrossRef]
17. Koomey, J.G.; Matthews, H.S.; Williams, E. Smart Everything: Will Intelligent Systems Reduce Resource Use? *Annu. Rev. Environ. Resour.* **2013**, *38*, 311–343. [CrossRef]

18. Arts, K.; Ioris, A.A.R.; Macleod, C.J.A.; Han, X.; Sripada, S.G.; Braga, J.R.Z.; van der Wal, R. Environmental Communication in the Information Age: Institutional Barriers and Opportunities in the Provision of River Data to the General Public. *Environ. Sci. Policy* **2016**, *55*, 47–53. [CrossRef]
19. Caragliu, A.; Nijkamp, P. Space and Knowledge Spillovers in European Regions: The Impact of Different Forms of Proximity on Spatial Knowledge Diffusion. *J. Econ. Geogr.* **2016**, *16*, 749–774. [CrossRef]
20. Wang, L.; Chen, L.; Li, Y. Digital Economy and Urban Low-Carbon Sustainable Development: The Role of Innovation Factor Mobility in China. *Environ. Sci. Pollut. Res.* **2022**, *29*, 48539–48557. [CrossRef]
21. Perez-Trujillo, M.; Lacalle-Calderon, M. The Impact of Knowledge Diffusion on Economic Growth across Countries. *World Dev.* **2020**, *132*, 104995. [CrossRef]
22. Yi, M.; Wang, Y.; Sheng, M.; Sharp, B.; Zhang, Y. Effects of Heterogeneous Technological Progress on Haze Pollution: Evidence from China. *Ecol. Econ.* **2020**, *169*, 106533. [CrossRef]
23. Zhou, Y.; Zhang, J.; Hu, S. Regression Analysis and Driving Force Model Building of CO_2 Emissions in China. *Sci. Rep.* **2021**, *11*, 6715. [CrossRef]
24. Horner, N.C.; Shehabi, A.; Azevedo, I.L. Known Unknowns: Indirect Energy Effects of Information and Communication Technology. *Environ. Res. Lett.* **2016**, *11*, 103001. [CrossRef]
25. Ren, S.; Hao, Y.; Xu, L.; Wu, H.; Ba, N. Digitalization and Energy: How Does Internet Development Affect China's Energy Consumption? *Energy Econ.* **2021**, *98*, 105220. [CrossRef]
26. Wu, H.; Xue, Y.; Hao, Y.; Ren, S. How Does Internet Development Affect Energy-Saving and Emission Reduction? Evidence from China. *Energy Econ.* **2021**, *103*, 105577. [CrossRef]
27. Shahbaz, M.; Wang, J.; Dong, K.; Zhao, J. The Impact of Digital Economy on Energy Transition across the Globe: The Mediating Role of Government Governance. *Renew. Sustain. Energy Rev.* **2022**, *166*, 112620. [CrossRef]
28. Yang, B.; Liu, B.; Peng, J.; Liu, X. The Impact of the Embedded Global Value Chain Position on Energy-Biased Technology Progress: Evidence from Chinas Manufacturing. *Technol. Soc.* **2022**, *71*, 102065. [CrossRef]
29. Gieraltowska, U.; Asyngier, R.; Nakonieczny, J.; Salahodjaev, R. Renewable Energy, Urbanization, and CO_2 Emissions: A Global Test. *Energies* **2022**, *15*, 3390. [CrossRef]
30. Capello, R.; Camagni, R. Beyond Optimal City Size: An Evaluation of Alternative Urban Growth Patterns. *Urban Stud.* **2000**, *37*, 1479–1496. [CrossRef]
31. Danish; Baloch, M.A.; Mahmood, N.; Zhang, J.W. Effect of Natural Resources, Renewable Energy and Economic Development on CO_2 Emissions in BRICS Countries. *Sci. Total Environ.* **2019**, *678*, 632–638. [CrossRef]
32. Ding, Y.; Yang, Q.; Cao, L. Examining the Impacts of Economic, Social, and Environmental Factors on the Relationship between Urbanization and CO_2 Emissions. *Energies* **2021**, *14*, 7430. [CrossRef]
33. Xu, Q.; Dong, Y.; Yang, R. Urbanization Impact on Carbon Emissions in the Pearl River Delta Region: Kuznets Curve Relationships. *J. Clean. Prod.* **2018**, *180*, 514–523. [CrossRef]
34. Notley, T. The Environmental Costs of the Global Digital Economy in Asia and the Urgent Need for Better Policy. *Media Int. Aust.* **2019**, *173*, 125–141. [CrossRef]
35. Collard, F.; Feve, P.; Portier, F. Electricity Consumption and ICT in the French Service Sector. *Energy Econ.* **2005**, *27*, 541–550. [CrossRef]
36. Longo, S.B.; York, R. How Does Information Communication Technology Affect Energy Use? *Hum. Ecol. Rev.* **2015**, *22*, 55–71. [CrossRef]
37. Sturgeon, T.J. Upgrading Strategies for the Digital Economy. *Glob. Strateg. J.* **2021**, *11*, 34–57. [CrossRef]
38. Gereffi, G. International Trade and Industrial Upgrading in the Apparel Commodity Chain. *J. Int. Econ.* **1999**, *48*, 37–70. [CrossRef]
39. Lu, H.; Peng, J.; Lu, X. Do Factor Market Distortions and Carbon Dioxide Emissions Distort Energy Industry Chain Technical Efficiency? A Heterogeneous Stochastic Frontier Analysis. *Energies* **2022**, *15*, 6154. [CrossRef]
40. Zhao, J.; Jiang, Q.; Dong, X.; Dong, K.; Jiang, H. How Does Industrial Structure Adjustment Reduce CO_2 Emissions? Spatial and Mediation Effects Analysis for China. *Energy Econ.* **2022**, *105*, 105704. [CrossRef]
41. Cardona, M.; Kretschmer, T.; Strobel, T. ICT and Productivity: Conclusions from the Empirical Literature. *Inf. Econ. Policy* **2013**, *25*, 109–125. [CrossRef]
42. Shi, K.; Yu, B.; Zhou, Y.; Chen, Y.; Yang, C.; Chen, Z.; Wu, J. Spatiotemporal Variations of CO_2 Emissions and Their Impact Factors in China: A Comparative Analysis between the Provincial and Prefectural Levels. *Appl. Energy* **2019**, *233*, 170–181. [CrossRef]
43. Benjamin, N.; Lin, B. Quantile Analysis of Carbon Emissions in China Metallurgy Industry. *J. Clean. Prod.* **2020**, *243*, 118534. [CrossRef]
44. Baron, R.M.; Kenny, D.A. The Moderator–Mediator Variable Distinction in Social Psychological Research: Conceptual, Strategic, and Statistical Considerations. *J. Personal. Soc. Psychol.* **1986**, *51*, 1173. [CrossRef]
45. Tone, K. A Slacks-Based Measure of Super-Efficiency in Data Envelopment Analysis. *Eur. J. Oper. Res.* **2002**, *143*, 32–41. [CrossRef]
46. Zhang, J.; Wu, G.; Zhang, J. The estimation of China's provincial capital stock: 1952–2000. *Econ. Res. J.* **2004**, *10*, 35–44.
47. Shan, Y.; Guan, Y.; Hang, Y.; Zheng, H.; Li, Y.; Guan, D.; Li, J.; Zhou, Y.; Li, L.; Hubacek, K. City-Level Emission Peak and Drivers in China. *Sci. Bull.* **2022**, *67*, 1910–1920. [CrossRef]
48. Zhao, T.; Zhang, Z.; Liang, S. Digital Economy, Entrepreneurship, and High-Quality Economic Development: Empirical Evidence from Urban China. *Manag. World* **2020**, *36*, 65–76.

49. Feng, Y.; Zhang, H. Network Effect, Demand Behavior and Market Size—Empirical Study and Policy Implications Based on the Express Induxtry. *China Ind. Econ.* **2021**, *1*, 115–135. [CrossRef]
50. Narayan, P.K.; Saboori, B.; Soleymani, A. Economic Growth and Carbon Emissions. *Econ. Model.* **2016**, *53*, 388–397. [CrossRef]
51. Guo, F.; Wang, J.; Wang, F.; Kong, T.; Zhang, X.; Cheng, Z. Measuring China's Digital Financial Inclusion: Index Compilation and Spatial Characteristics. *China Econ. Q.* **2020**, *19*, 1401–1418.
52. Xiao, H.; Zhou, Y.; Zhang, N.; Wang, D.; Shan, Y.; Ren, J. CO_2 Emission Reduction Potential in China from Combined Effects of Structural Adjustment of Economy and Efficiency Improvement. *Resour. Conserv. Recycl.* **2021**, *174*, 105760. [CrossRef]
53. Chen, Z.; Wu, S.; Ma, W.; Liu, X.; Cai, B.; Liu, J.; Jia, X.; Zhang, M.; Chen, Y.; Xu, L. Driving Forces of Carbon Dioxide Emission for China's Cities: Empirical Analysis Based on Extended STIRPAT Model. *China Popul. Resour. Environ.* **2018**, *28*, 45–54.
54. Glaeser, E.L.; Kahn, M.E. The Greenness of Cities: Carbon Dioxide Emissions and Urban Development. *J. Urban Econ.* **2010**, *67*, 404–418. [CrossRef]
55. Chen, S.; Chen, D. Air Pollution, Government Regulations and High-Quality Economic Development. *Econ. Res. J.* **2018**, *53*, 20–34.
56. Zhang, C.; Zhou, X. Does Foreign Direct Investment Lead to Lower CO_2 Emissions? Evidence from a Regional Analysis in China. *Renew. Sustain. Energy Rev.* **2016**, *58*, 943–951. [CrossRef]
57. Apergis, N.; Pinar, M.; Unlu, E. How Do Foreign Direct Investment Flows Affect Carbon Emissions in BRICS Countries? Revisiting the Pollution Haven Hypothesis Using Bilateral FDI Flows from OECD to BRICS Countries. *Environ. Sci. Pollut. Res.* **2022**, 1–13. [CrossRef]
58. Acheampong, A.O.; Amponsah, M.; Boateng, E. Does Financial Development Mitigate Carbon Emissions? Evidence from Heterogeneous Financial Economies. *Energy Econ.* **2020**, *88*, 104768. [CrossRef]
59. Wang, L.; Zhao, Z.; Xue, X.; Wang, Y. Spillover Effects of Railway and Road on CO_2 Emission in China: A Spatiotemporal Analysis. *J. Clean. Prod.* **2019**, *234*, 797–809. [CrossRef]
60. Zhang, Y.-J. The Impact of Financial Development on Carbon Emissions: An Empirical Analysis in China. *Energy Policy* **2011**, *39*, 2197–2203. [CrossRef]
61. Khan, S.; Peng, Z.; Li, Y. Energy Consumption, Environmental Degradation, Economic Growth and Financial Development in Globe: Dynamic Simultaneous Equations Panel Analysis. *Energy Rep.* **2019**, *5*, 1089–1102. [CrossRef]
62. Stock, J.; Yogo, M. Asymptotic Distributions of Instrumental Variables Statistics with Many Instruments. *Identif. Inference Econom. Model. Essays Honor. Thomas Rothenberg* **2005**, *6*, 109–120. [CrossRef]
63. Zhang, X.; Wan, G.; Jiajia, Z.; Zongyue, H. Digital Economy, Financial Inclusion and Inclusive Growth. *China Econ.* **2020**, *15*, 92–105.
64. Zhang, J.; Jiang, H.; Liu, G.; Zeng, W. A Study on the Contribution of Industrial Restructuring to Reduction of Carbon Emissions in China during the Five Five Year Plan Periods. *J. Clean. Prod.* **2018**, *176*, 629–635. [CrossRef]

energies

MDPI

Article

Disruptive Displacement: The Impacts of Industrial Robots on the Energy Industry's International Division of Labor from a Technological Complexity View

Weiming Zhang [1], Jiachao Peng [1,2,*] and Lian Zhang [3]

1 School of Law & Business, Wuhan Institute of Technology, Wuhan 430205, China; 22108010010@stu.wit.edu.cn
2 Center for High Quality Collaborative Development of Resources, Environment and Economy, Wuhan Institute of Technology, Wuhan 430205, China
3 School of Art & Design, Wuhan Institute of Technology, Wuhan 430205, China; zhanglian@cug.edu.cn
* Correspondence: jiachao.peng@wit.edu.cn

Abstract: In light of the growing economic uncertainties worldwide, the use of industrial robots has emerged as a significant opportunity for improving the production efficiency and the international division of labor in China's energy industry. This study employed a two-way fixed-effect model utilizing data from 31 Chinese provinces between 2011 and 2019 to investigate the impact of industrial robots on the energy industry's participation in the international division of labor. The results of the study indicated that the widespread application of industrial robots can boost the international division of labor status of China's energy sector. This conclusion remains robust even after addressing the potential endogeneity issues and conducting a range of sensitivity tests. Furthermore, our findings suggest that the regions that possess abundant energy resources or exhibit a lower carbon intensity are more likely to leverage the use of industrial robots to increase the technological sophistication and enhance their participation in the international division of labor. The application of industrial robots in the energy industry can enhance the international division of labor through two distinct channels: optimizing the factor structure and reducing the export costs. Our findings have important policy implications for ensuring energy security and improving the energy industry's participation in the international division of labor.

Keywords: industrial robot; international division of labor; export technology complexity

Citation: Zhang, W.; Peng, J.; Zhang, L. Disruptive Displacement: The Impacts of Industrial Robots on the Energy Industry's International Division of Labor from a Technological Complexity View. *Energies* **2023**, *16*, 3349. https://doi.org/10.3390/en16083349

Academic Editor: Michael Short

Received: 31 January 2023
Revised: 30 March 2023
Accepted: 6 April 2023
Published: 10 April 2023

1. Introduction

Since the 1970s, China has developed a processing model that leverages its advantages in the international division of labor, resulting in a significant expansion of trade. By 2013, China had become the world's largest goods trading nation, with its goods trade exceeding RMB 31 trillion by 2019. International trade has been a critical driver of China's economic development [1]. However, an undeniable fact is that the gap between China's energy supply and demand is expanding, and this poses a significant challenge to its economic development. China's external dependence on crude oil and natural gas reached 72% and 46%, respectively, in 2021. The slow progress in optimizing China's energy consumption and import structure and the increasing concentration of resources are exacerbating China's energy security problem [2]. Moreover, the global economic situation has had a profound impact on the global value chain of the division of labor system. In recent years, the global environment has become increasingly complex, and protectionism is on the rise. Global economic development has reached an important crossroads, and multilateralism faces greater uncertainty. The future of free trade mechanisms and the international division of labor is uncertain. Against this backdrop, China's need for effective energy security is facing greater challenges [3]. To respond to the challenges posed by the changing world economic environment, the 2021 National Work Conference on Development and Reform highlighted the need to strengthen the construction of energy production, supply, storage, and marketing systems to ensure energy security. Among these measures, enhancing high-quality energy trade, breaking the strong position of multinational companies in the international division of labor, promoting Chinese energy companies to deepen their involvement in the value chain [4], and improving their position in the international division of labor system have become urgent issues that need to be addressed.

According to recent research, technological advancements can significantly impact an individual's level of participation in the global value chains' division of labor. A firm's total factor productivity largely influences its position in the global production chain. Increasing the income from technology and improving the total factor productivity can improve China's unfavorable position in related industries in the global value chain. Moreover, enhancing the research and development (R&D) intensity and rationally applying technological achievements can also improve the industry's international division of labor position [5,6]. However, the implementation of protectionist policies in developed countries, where the protection of relevant technologies is prioritized, poses a challenge for upgrading the technology and international division of labor status in developing countries. China's changing national conditions, such as an aging population, rising labor costs, and the uncertainty surrounding foreign demand, also pose significant challenges for product upgrading. The development and progress of artificial intelligence technology in the 21st century have propelled the robust development of modern industrial robots, which have brought together various modern high technologies, significantly contributing to an enterprise's productivity [7]. This presents new opportunities and possibilities for overcoming China's current dilemma and improving the energy sector's international division of labor. This is especially crucial in the context of China's "carbon neutrality" efforts, which demand higher development standards for the energy industry to enhance its quality and efficiency. Therefore, it is of great practical significance to investigate the impact of industrial robots on the international division of labor in the energy industry.

The application of industrial robots has an impact on the structure of the workforce. Although scholarly research on the subject has yet to reach a fully consistent conclusion, it represents a departure from previous technological advances. Some scholars contend that the adoption of industrial robots will increase the demand for skilled workers while reducing the need for unskilled workers and simultaneously leading to an increase in the use of skilled personnel [8]. However, others disagree, arguing that industrial robots differ from other technological advances in that their impact may cover the workforce at different skill levels [5,6,9]. The manner in which the adoption of industrial robots affects the structure of the workforce is primarily influenced by the substitution effect and the productivity effect. The substitution effect refers to the way that industrial robots can occupy positions that would otherwise be held by human workers and perform the necessary tasks. The productivity effect refers to the fact that technological advances can increase productivity while driving the need for related jobs that are still in the technological stagnation stage. This increase in productivity reduces production costs and product prices, resulting in an increase in real income for the population and the related demand for consumer goods and services. Ultimately, this raises the labor demand for tasks that have not been replaced by related technological advances [6,10].

The economic impact of industrial robots is the second factor to consider. Extensive research conducted by national and international scholars demonstrated that industrial robots are a primary tool for enhancing productivity. A panel data study that examined the use of industrial robots concluded that their application led to a decrease in product prices and contributed to an increase in the total factor productivity [5]. In a study conducted by Chen et al. (2019) [11], it was found that the utilization of artificial intelligence resulted in a reduction in the labor demand during the production processes, an increase in the total factor productivity, an acceleration of the accumulation of capital, and a consequent increase in the return on capital. These factors positively impacted economic growth and served to mitigate the effects of aging on economic growth [11]. Another study conducted by Lin et al. (2020) [12] employed a dynamic general equilibrium model that combined AI and heterogeneous capital to examine their impact on economic growth. The results indicated that AI played a significant role in optimizing the capital structure, driving economic growth, and promoting an increase in the population's consumption level [13].

In recent years, industrial robotics have undergone increasing advancements and their applications have expanded to various fields, prompting more scholars to study the relationship between their use and international trade. Goldfarb and Trefler (2018) [14] were the first to explore the relationship between AI and international trade, revealing that factors such as economies of scale, knowledge creation, and the geographical location of knowledge diffusion may contribute to the impact of AI on international trade patterns. Since then, the adoption of industrial robotics has been shown to affect the status of neighboring countries in the international division of labor, as demonstrated by Artuc et al. (2020) [15] who found that a large-scale implementation of robots in the North can lead to increased imports in the South, resulting in both regions achieving a higher production and trade of intermediate and final goods. However, the expansion of production and trade of intermediate goods in the North may come at the expense of a smaller share for the South, potentially hindering the growth of the international division of labor in the region. Scholars have also noted that non-manufacturing industries have shown a greater interest in the use of industrial robots and related technologies [16]. The application of industrial robots and the digitization in the manufacturing industry has also led to an improvement in the international division of labor status of service industries and the quality of service trade development [17]. Empirical studies have further demonstrated that industrial robots have different technological levels in the services industry [10,18,19].

The research on industrial robots is continuously expanding, with scholars increasingly exploring their implications for international trade. Some scholars have even suggested that industrial robots could provide a vital opportunity for developing countries to overcome their challenges and achieve a competitive advantage. However, much of the current research on industrial robots and their impact on global value chains and the international division of labor has been concentrated on the manufacturing and services sectors. As such, there is a gap in the literature that examines the influence of industrial robots on the exports within the energy sector, creating an opportunity for this paper to contribute to the field.

Compared to the existing literature, this paper's potential contributions are threefold. (1) The research questions focus not only on the impact of industrial robots on labor and employment, but also on the actual influence of the robot applications on industry progress as artificial intelligence technology continues to advance. (2) The research content delves into the microscopic mechanisms behind the impact of industrial robots on the energy industry's participation in the international division of labor. This paper not only examines the effect of industrial robots on the energy industry's participation in the international division of labor but also analyzes the path of this impact in depth. (3) The research dataset was constructed by manually matching the industrial robot data, industrial enterprise database data, and customs database data, and providing a sample that can be used to study the technical complexity of the industrial robots affecting the energy exports. This paper confirms that industrial robots play a role in enhancing the technical complexity of energy product exports, and further verifies that they can optimize the factor allocation and reduce the export costs, providing a theoretical basis for leveraging industrial robots to enhance the international division of labor in China's energy industry. Drawing on existing research, this paper constructed a panel dataset using robot data from the International Robot Federation (IRF) website for China and other countries, as well as provincial-level data from databases, such as EPS. The dataset was used to investigate the impact of the industrial robot applications on China's international division of labor position in the energy industry and to examine the causal pathways of this impact. The plausibility of the results was tested for the period spanning from 2011 through 2019.

The following is the proposed sequencing of the paper. In Section 2, we will describe the theoretical mechanism analysis and hypotheses that underpin this study. In Section 3, we will explain the selection of the variables and the model construction adopted for this study. In Section 4, we will provide an empirical analysis of the theoretical framework developed in Section 2. Section 5 will provide further tests of the influence mechanism, while Section 6 will provide the conclusion and policy implications of this study.

2. Theoretical Analysis and Research Hypothesis

2.1. The Application of Industrial Robots Affects the Complexity of Export Technology

Robots are a highly sophisticated technology that have been increasingly adopted in various stages of corporate product development and production processing. As such, robots can exert a significant impact on the progress of various industries.

Product innovation is widely recognized as a critical driver of export complexity [20,21]. In the process of product development, innovation is often a trial and error process, requiring significant R&D efforts and experimentation, which can lead to increased marginal costs [22]. In contrast, robots offer a solution to the challenges of R&D by providing rapid and precise results, allowing for more reliable and efficient experimentation. This not only improves the quality of the product but also shortens the development cycle, enabling enterprises to introduce competitive products into the market quickly. By increasing the technical input during the export product development process, product innovation is an effective means to expand the product coverage and optimize the product quality [23], ultimately leading to an increase in the export technology complexity.

The deployment of industrial robots in the production and processing system holds the potential to transform the original production process and increase the production efficiency. One advantage of robots is their ability to assess product states quickly and accurately and autonomously perform the processing using hardware, such as light and sound sensing, and software, such as big data. As a result, the application of robots can improve the product quality of enterprises and enhance the technical complexity of exported products. However, the implementation of robots may also lead to the replacement of some labor. Robots can perform tasks in the production process that are impossible for human labor, resulting in a possible shift in the production process and an increase in labor productivity. The scale effect generated from the increased labor productivity enables industries to exploit their differentiation advantages and boost the technological complexity of their exports. Simultaneously, the enhancement of the labor productivity implies that enterprises can regulate their production costs, and effective cost management is a crucial means for enterprises to enhance their export complexity [24–26].

In summary, the utilization of robots in critical aspects of enterprise product development, production, and processing presents an opportunity to enhance the export complexity of the industry. Building upon this insight, we propose the following.

Hypothesis 1. *Industrial robot applications promote the upgradation of the technological complexity of exports in the energy industry.*

2.2. The Mechanism of Industrial Robots Affecting the Technical Complexity of Exports

More precisely, the integration of industrial robots advances the technological complexity of enterprise exports by the following two means.

Optimizing the factor structure is a crucial pathway for industrial robots to upgrade the export technical complexity. According to the factor substitution theory, the capital–labor ratio is closely linked to the usage of capital and labor. Specifically, if the price of a production factor increases, technology will progress towards reducing the use of that factor and eventually replacing it, leading to biased technological progress [27]. Technological advancements, such as industrial robots, affect both capital (K) and labor (L), thereby changing the factor allocation structure of the firms. With the changing national conditions, more rational enterprises will further invest in capital to maximize profits. Industrial robots are primarily a result of capital deepening, capable of filling low- and medium-skilled labor positions and completing the corresponding work. By scaling up the usage of industrial robots and other technologies, enterprises can reduce the usage of labor (L) and increase capital investment (K), thereby enhancing the capital–labor ratio (K/L), accelerating capital deepening, and enabling the optimization and upgrading of the factor structure. Factor markets are also linked to industrial robots. The existing literature suggests that encouraging companies to engage in intelligent production and increasing the application of industrial robots is an effective approach to optimize the production costs when the price of capital decreases [28]. The application of industrial robots can alter the proportion of the factors invested in production and effectively manage the production costs of the industry, thereby fulfilling the need for the increased technological complexity of the exports in the energy sector.

Furthermore, the application of industrial robots plays a significant role in reducing the cost of exporting products, thereby enhancing the technical complexity of the industry's exports. On the one hand, robots facilitate the synergistic development of various aspects of international trade, including transportation, storage, packaging, loading, and unloading. This results in reduced expenses during transportation and distribution and meets the demand of the enterprises for low transportation costs. On the other hand, the practical application of robot technology, such as the sorting and handling of robots, allows for intelligent product storage, leading to a further decrease in the overall product storage costs in the industry. As a result, whether in logistics and transportation or intelligent storage, the widespread usage of robots has led to a reduction in the fixed costs of exports, enabling enterprises to invest more capital in research and development, enhance the technical content of their products, and improve the technical complexity of their exports. Therefore, based on these observations, we propose a second hypothesis for this study.

Hypothesis 2. *The optimization of the factor allocation and reduction in export costs through the application of industrial robots promotes an increase in the technical complexity of the exports.*

3. Empirical Study Design

3.1. Indicator Selection

3.1.1. Explained Variables

The explained variable in this paper is the technological complexity of the exports in China's energy industry ($Complx$). The concept of the export technological complexity was first proposed by Hausmann et al. (2007) [25]. Since then, various methods for measuring this variable have been proposed by scholars worldwide. Xu and Lu (2009) [29] modified the calculation method by utilizing provincial export data and GDP per capita when cross-country comparisons were not necessary. In this study, we calculated the export comparative advantage index of the different energy products in each province and used it as the weighted average to obtain the technological complexity of the exported energy products ($PRODY_n$).

$$PRODY_n = \sum_m \frac{\frac{X_{mn}}{X_m}}{\sum_m \frac{X_{mn}}{X_m}} Y_m \quad (1)$$

In this study, the subscripts m and n represent the provinces and products, respectively. X_{mn} represents the export value of product n in province m, while X_m represents the total export value of all the products in province m. Additionally, Y_m represents the GDP per capita of province m.

After calculating the technological complexity of the exported energy products, we then used the export weight of each province to derive the export technical complexity of each province ($EXPY_m$). This was calculated using the following formula.

$$EXPY_m = \sum_n \frac{X_{mn}}{X_m} PRODY_n \quad (2)$$

In this paper, we considered the requirements of the panel data regression and data availability and classified the energy sources into six categories, namely coal, coke or semi-coke, crude oil, oil, natural gas, and electricity. This classification was based on the data obtained from the various sources, such as the China Energy Statistical Yearbook, Chinese customs data, and the EPS micro database, which provided the data related to the energy industry. Using this information, we calculated the technical complexity of the exports in the energy industry.

3.1.2. Core Explanatory Variables

In this paper, the level of industrial robot adoption was measured by the density of industrial robot use, which was represented by the core explanatory variable, the industrial robot penetration (*Robots*). Following the methods proposed by Acemoglu and Restrepo (2020) [5] and Wang and Dong (2020) [30], this paper used the industry-level robot penetration and the labor employment ratio of each firm to calculate the level of industrial robot adoption by each firm. The specific measurement formula is presented as follows.

$$Robots_{jit} = \frac{PWP_{jit=2011}}{ManuPWP_{t=2011}} \times \frac{MR_{it}^{CH}}{L_{i,t=2011}^{CH}} \quad (3)$$

Specifically, the variable $Robots_{jit}$ measures the penetration of industrial robots in industry i and enterprise j in year t. It was calculated as the product of three terms: (1) the ratio of the proportion of employees in the production department of enterprise j in industry i in the manufacturing industry in 2011 (base period) to the median proportion of employees in the production department of all the enterprises in the manufacturing industry in 2011 ($\frac{PWP_{jit=2011}}{ManuPWP_{t=2011}}(t = 2011)$), (2) the stock of industrial robots belonging to the enterprise j in industry i in year t (MR_{it}^{CH}), and (3) the employment number of industry i in China in 2011 ($L_{i,t=2011}^{CH}$). Finally, the calculated industrial robot penetration of each listed enterprise was matched and summed with the provinces one by one to obtain the penetration of industrial robots at the provincial level.

3.1.3. Control Variables

To examine the impact of industrial robots on the technological complexity of the exports in the energy industry and to ensure the robustness and reliability of our econometric regression results, this paper controlled the other variables that might have affected the technological complexity of the energy exports (see Table 1), following the approach of Xu et al. (2022) [31].

Table 1. Variable definitions.

	Variable Symbolic	Variable Name	Variable Definition
Explained variable	*Complx*	Export technology complexity	Technical content of export products
Core explanatory variable	*Robots*	Industrial robot penetration	Robot stock/employment
Control variable	*Develop*	Economic development level	Household consumption level index
	FDI	Foreign direct investment	Total registered foreign investment
	Patent	Technology innovation level	Patent application authorization number

Table 1. *Cont.*

	Variable Symbolic	Variable Name	Variable Definition
Control variable	*Labor*	Human capital	Average number of students in colleges and universities
	Index	Internet development	Internet broadband access port
	Fin	Financial development	Science and technology finance index

In this study, we examined the factors affecting the application of industrial robots by the enterprises. Specifically, the level of economic development, as measured by the level of residential consumption, played a crucial role in determining the ability of a region to utilize high technology. We used the level of residential consumption in each province to represent this variable [32].

Furthermore, foreign direct investment (FDI) can crowd the market when introducing technology, thereby affecting the technological sophistication of the exports in the energy industry [10,13,33]. To measure FDI, we used the total amount of foreign registered investment in each province.

The level of technological innovation, as represented by the number of patent applications granted in each province, can significantly impact the export complexity of a country's products, which in turn changes the export structure. Therefore, we used this variable to indicate the level of technological innovation [34].

Human capital is essential for providing labor to the production of enterprises. The quality of human capital affects the R&D capability of enterprises and determines the technical complexity of their products for export. In this study, we used the average number of students per 100,000 in higher education in each province to represent this variable [26].

Internet development strengthens inter-industry linkages and facilitates the division of labor in the industry. Moreover, it improves the efficiency of resource utilization and the flow of technology. To measure the internet development, we used the number of internet broadband access ports in each province [35].

Lastly, we examined the degree of financial development, which can help alleviate information asymmetry, improve the efficiency of fund utilization, and encourage the industry to enhance the sophistication of export technology. As an essential means of combining finance and innovation, technology finance reflects the goal of financial development. Therefore, we used the Technology Finance Index to represent this variable [36].

3.2. Basic Model Settings

In this paper, we constructed a basic econometric model by identifying the indicators to measure the consumption and industrial upgrading.

$$Complx_{it} = \alpha_0 + \alpha_1 Robots_{it} + \alpha_2 X_{it} + \mu_i + \gamma_t + \varepsilon_{it} \tag{4}$$

where subscripts i and t represent the provinces and years, respectively; $Complx_{it}$ represents the technical complexity of the energy export of province i in year t, which is the explained variable of this paper; $Robots_{it}$ represents the consumption scale and consumption structure of province i in year t, which is the core explanatory variable of this paper; X_{it} is the control variable; μ_i and γ_t represent the unobserved provinces and the time fixed effects, respectively; ε_{it} is a random disturbance term; and α is the main parameter. In Equation (4), the core estimation parameter is represented by α_1. A significantly positive value of α_1 indicates that the increased use of industrial robots promotes the improvement of the export technology complexity in China's energy industry. On the other hand, if the value of α_1 is not significant, the conclusion cannot be supported.

3.3. Data Source and Variable Description

This paper aims to investigate the impact of industrial robots on the technical complexity of China's energy industry exports using the panel data from 31 provinces in China over a period of 9 years, from 2011 to 2019. The data sources used in this study mainly included customs export data, statistical yearbooks, and the EPS database of each province. To avoid the influence of outliers and potential bias in the conclusions, a 1% tail reduction was applied to the collated data before and after conducting the further regression analysis on the processed data. The descriptive statistics of each variable are presented in Table 2.

Table 2. Descriptive statistics.

Var. Name	Obs.	Mean	SD	Min	Median	Max
Complx	279	16.657	90.456	−97.627	−1.704	1211.009
Robots	279	1401.858	2387.037	16.770	553.219	17,363.777
Develop	279	102.526	1.226	100.567	102.266	106.338
FDI	279	1662.923	2744.061	7.259	621.000	19,533.000
Patent	279	50,471.229	75,386.375	121.000	22,820.000	5.27×10^5
Labor	279	2557.991	802.073	1082.149	2383.000	5612.870
Index	279	202.348	91.647	16.220	212.360	410.280
Fin	279	806.145	724.299	7.600	600.329	4112.233

Note: "Obs." indicates observation; "Mean" indicates mean; "SD" indicates standard deviation; "Min" indicates minimum; "Median" indicates median; "Max" indicates maximum.

Table 2 shows that the technical complexity of the energy industry exports (*Complx*) ranged from −97.627 to 1211.009, indicating a significant variation in the level of the technical complexity across the different provinces. Similarly, the industrial robot penetration (*Robots*) varied widely, with a minimum value of 16.770 and a maximum value of 17,363.777, suggesting a significant disparity in the adoption of industrial robots across the different regions of China.

4. Results

The estimation methods for the short panels (N large T small) included three types of mixed regressions, fixed effects, and random effects models. To determine the appropriate regression model, the Hausman test was applied. The test results indicated that the χ^2 statistic was 27.15 with a p-value of 0.0001, which rejects the random effect at the 1% level. Thus, the fixed-effect model should be used.

4.1. Data Source and Variable Description

In this study, we initially analyzed the effect of industrial robots on the technical complexity of the energy exports by regressing Equation (1) using a two-way fixed-effect model that was controlled for the province and year effects.

The results of the baseline regression are presented in Table 3. Columns (1) and (2) display the results without the inclusion of the region-fixed and year-fixed effects, while columns (3) and (4) show the control for both the region and year-fixed effects. As demonstrated in Table 3, the estimated coefficient of the industrial robot penetration remained significantly positive at the 5% statistical level, regardless of the inclusion of the region and year-fixed effects. This finding aligned with our prior expectations and suggests that the use of industrial robots has a noteworthy positive impact on the technical complexity of the energy exports. Furthermore, the application of industrial robots can enhance the competitiveness of the export products within the energy industry and foster international improvements in the division of labor.

Table 3. Regression results of the two-way fixed-effect model.

	(1)	(2)	(3)	(4)
	Complx	*Complx*	*Complx*	*Complx*
Robots	0.00449 *** (0.00109)	0.00820 *** (0.00246)	0.00399 ** (0.00145)	0.00615 * (0.00292)
Develop		3.524 (2.005)		−1.157 (4.742)
FDI		−0.00689 ** (0.00222)		−0.00863 *** (0.00250)
Patent		0.000546 *** (0.000111)		0.000740 *** (0.000176)
Labor		0.0175 ** (0.00622)		−0.0330 * (0.0131)
Index		0.0701 (0.0376)		1.017 ** (0.322)
Fin		−0.0433 *** (0.00950)		−0.0401 ** (0.0121)
_cons		−399.7 (208.6)		170.5 (501.7)
Id effect	No	No	Yes	Yes
Year effect	No	No	Yes	Yes
N	279	279	279	279
R^2			0.080	0.254

Note: The numbers in the brackets are standard errors; ***, **, and * indicate a statistical significance at the 1%, 5%, and 10% levels, respectively. "No" means that the effect was not controlled; "Yes" means that the effect was controlled.

Column (4) of the regression results in Table 3 demonstrated that a 1% increase in the robot penetration led to a 0.162% increase in the technical complexity of the energy industry exports, assuming the other control variables remained constant. This finding confirmed Hypothesis 1 and underscored the crucial role that industrial robots play in promoting the international division of labor within the energy industry. Notably, among the control variables, the coefficients for the level of technological innovation and the degree of internet development were significantly positive at the 1% level, indicating that these factors facilitated the improvement of the technical complexity of the energy exports, which was in line with the previous research. Conversely, FDI was significantly negative at the 1% level, potentially due to the market competition and inadequate technology spillover from foreign capital. Labor was also significantly negative at the 10% level, implying that merely increasing the quantity of labor in China's energy sector is insufficient to meet the demands of the technological sophistication, highlighting the need for quality talent. Additionally, the coefficient of the degree of financial development was significantly negative, potentially due to a mismatch of financial resources in China's energy industry, which hindered the overall level of the international division of labor.

4.2. Endogenetic Treatment

4.2.1. Lagging the Core Explanatory Variables by One Period

Since the introduction and use of industrial robots in production processes may involve a lag effect, companies may not immediately achieve the desired technological upgrade when implementing them. Instead, they might gradually achieve it in subsequent periods. To account for this possibility, this section introduces the robot penetration index with a lag of one period into the model to test the impact of the industrial robot penetration on the technological sophistication of the exports. The results in column (1) of Table 4 demonstrated that the robot penetration at the one-period lag remained significantly positive at the 1% level, further supporting the robustness of Hypothesis 1.

Table 4. Endogenous treatment.

	(1)	(2)	(3)	(4)	(5)
	Complx	*Robots*	*Complx*	*Robots*	*Complx*
L.Robots	0.00831 * (0.00342)				
Robots_US		0.591 *** (4.771)			
Robots			0.037 ** (2.433)		
Robots_JP				−0.262 *** (−3.751)	
Robots					0.043 ** (2.509)
rk LM test		0.000		0.000	
rk F test		264.35		255.38	
Control variables	YES	YES	YES	YES	YES
Id effect	YES	YES	YES	YES	YES
Year effect	YES	YES	YES	YES	YES
Observations		279	279	279	279
R-squared		0.872	0.140	0.868	0.106

Note: The numbers in the brackets are standard errors; ***, **, and * indicate a statistical significance at the 1%, 5%, and 10% levels, respectively. "No" means that the effect was not controlled; "Yes" means that the effect was controlled.

4.2.2. Two-Stage Least Squares (2SLS) for the Tool Variable Estimation

Endogeneity problems may arise when regressing the industrial robot penetration in each province on the technological complexity of the exports in the energy sector, as an increased level in the international division of labor in energy could affect the industrial robot penetration. Firstly, a region's higher level in the energy division of labor could result from its strong research capability and better layout of high-tech industries, which would make it easier and less costly to receive and use industrial robots in the region. This could lead to a reverse causality between the explanatory variables and the core explanatory variables. Secondly, different companies may develop their robot use programs according to their actual situation and the requirements of their development direction, based on their current division of labor status. To address this endogeneity issue, we measured the industrial robot use as an instrumental variable, as suggested by Acemoglu and Restrepo (2020). They found that the competition among large manufacturing countries is intense, and the technologies and equipment used in the competition are similar. Therefore, the industrial robot use can be used as a valid instrumental variable when studying the impact of the industrial robot use on the employment in the United States.

To mitigate the potential endogeneity problems affecting the findings of this paper, this section followed the approach of Yan et al. (2021) [37] and Wang and Dong (2020) [30] by incorporating the stock of industrial robots from the United States and Japan in Equation (3) instead of relying solely on the Chinese industrial robot stock. Additionally, the newly computed industrial robot penetration at the provincial level in China was employed as an instrumental variable, and the two-stage least squares (2SLS) was used for the instrumental variable estimation. The data on the stock of industrial robots in the United States and Japan were sourced from the IFR. By employing this methodology, this paper can better address the potential endogeneity concerns and provide more robust and reliable results.

The regression results are presented in Table 4, where columns (2) and (4) present the results of the one-stage regression, and columns (3) and (5) show the results of the regression using the penetration degrees calculated as the instrumental variables from the industrial robot stock in the United States (*Robots_US*) and Japan (*Robots_JP*), respectively. To address the endogeneity issue, this paper used the newly calculated industrial robot penetration at the provincial level in China as an instrumental variable and employed the two-stage least squares estimation. The results in Table 4 showed that the application of industrial robots was significantly positive with the export technological complexity, which was consistent with the previous benchmark regressions, indicating that the application of industrial robots still had an impact on the export technological complexity after the endogeneity was addressed. It is worth mentioning that in this result, the influence of the instrumental variables on the export technical complexity was greater than that of the Chinese industrial robot penetration. This may be attributed to the earlier industrialization and stronger technological strength of the United States and Japan. The use of industrial robots enables different industries to cooperate with each other, improving the efficiency of the industrial robot use. The rk F-statistic was 264.35 and 255.38 with *Robots_US* and *Robots_JP* as instrumental variables, respectively, which were greater than the critical values, indicating that no weak instrumental variable problem was present. The p-values for both the rk LM tests were 0, rejecting the original hypothesis, and there was no under-identification problem.

4.3. Robustness Test

4.3.1. Add Control Variable

The technological complexity of the energy exports was influenced by multiple factors, including the infrastructure and firm factors. Therefore, the control-related factors may overlook other important factors and lead to non-robust estimation results. To account for this, we included additional control variables to control for the effects of the other factors. Specifically, we examined the level of infrastructure improvement (inf) by using road area per capita (square meters) as an indicator. The influence of the firm factor was examined by selecting the share of employees of state-owned enterprises in the energy sector as a variable (*property*).

Table 5 presents the estimation results after adding the two control variables mentioned above in Column (1). The results show that the regression coefficient of the industrial robot penetration remained significantly positive even after accounting for the influence of the other factors on the technological complexity of the exports in the energy sector at different levels. This suggests that the influence of the industrial robot penetration on the technological complexity of the energy exports was not affected by the other factors, and that the previous estimation results were robust.

Table 5. Robustness test.

	(1)	(2)
	Complx	***Complx***
Robots	0.00603 * (0.00296)	0.00000471 * (0.00000223)
Develop	−1.470 (4.814)	−0.000934 (0.00358)
FDI	−0.00877 *** (0.00252)	−0.00000660 *** (0.00000191)

Table 5. *Cont.*

	(1)	(2)
	Complx	*Complx*
Patent	0.000735 *** (0.000178)	0.000000565 *** (0.000000134)
Labor	−0.0292 * (0.0137)	−0.0000252 * (0.0000100)
Index	1.000 ** (0.327)	0.000776 ** (0.000246)
Fin	−0.0402 ** (0.0122)	−0.0000307 ** (0.00000924)
Inf	−1.227 (1.262)	0.210 (0.379)
Property	3.108 (21.18)	
_cons	210.3 (511.9)	
Id effect	YES	YES
Year effect	YES	YES
N	279	279
R^2	0.257	0.254

Note: The numbers in the brackets are standard errors; ***, **, and * indicate a statistical significance at the 1%, 5%, and 10% levels, respectively. "No" means that the effect was not controlled; "Yes" means that the effect was controlled.

4.3.2. Replace Dependent Variable

Equation (1) suggests that as the world per capita income increases, the technical complexity of the exported products tends to increase, while the characteristics of the products generally remain stable over time. To ensure the intertemporal stability of the product characteristics, a standard technical complexity index for the exported products was introduced [30]. The formula for the standard technical complexity index is as follows.

$$PRODY_n = \frac{PRODY_n - PRODY_{min}}{PRODY_{max} - PRODY_{min}} \quad (5)$$

The intertemporal stability of the product characteristics was ensured by introducing the standard technical complexity index of the exported products, as shown in Equation (5). Here, $PRODY_{min}$ and $PRODY_{max}$ represent the minimum and maximum product technical complexity of all the export products, respectively, and $PRODY_n$ represents the technical complexity of the export product standards. The estimated results in column (2) of Table 5 did not differ significantly from the benchmark results, indicating that the inclusion of the standard technical complexity index did not affect the robustness of the previous estimation results.

4.4. Heterogeneity Test

4.4.1. Distinguish Energy Output

The regional differences in the energy production and the industrial robot application may lead to varying impacts from the use of industrial robots on the technical complexity of the energy exports. In order to account for this, we adopted a group regression approach to analyze the samples of the regions that were rich and not rich in energy production. The results, presented in columns (1) and (2) of Table 6, show that the use of industrial robots in regions with abundant energy production significantly increased the technical complexity of the energy exports, while the effect of industrial robots in regions with a more backward energy production was not significant. This difference could be attributed to the fact that the regions with a low energy production may not be able to use industrial robots on a large scale due to the region's actual situation, and thus, are unable to achieve a scale effect. Additionally, the regions with an abundant energy production were more likely to have access to targeted financial and technical support, leading to higher levels of industrial robot adoption. As a result, the factor structure optimization and export cost reduction from the application of industrial robots were more pronounced in the energy production-rich regions, which was conducive to the improvement of the technical complexity of the energy industry exports in these regions.

Table 6. Heterogeneity test.

	(1)	(2)	(3)	(4)
	Complx	*Complx*	*Complx*	*Complx*
Robots	0.0220 *** (0.00316)	0.000949 (0.00943)	0.00149 (0.00303)	0.0180 ** (0.00617)
Develop	−15.48 * (6.940)	−2.413 (5.487)	−0.944 (1.200)	−7.477 (12.75)
FDI	−0.0139 *** (0.00346)	−0.000968 (0.00624)	−0.0190 *** (0.00221)	−0.0141 * (0.00589)
Patent	0.0000990 (0.000214)	0.000364 (0.000348)	−0.000134 (0.000191)	0.000378 (0.000309)
Labor	0.0128 (0.0186)	−0.0293 (0.0164)	0.00318 (0.00435)	−0.0651 * (0.0313)
Index	−0.0261 (0.441)	1.063 * (0.412)	0.157 (0.102)	2.426 ** (0.850)
Fin	−0.0402 ** (0.0151)	0.0273 (0.0251)	0.0150 * (0.00649)	−0.0570 * (0.0269)
_cons	0.0220 *** (0.00316)	0.000949 (0.00943)	96.72 (128.2)	876.4 (1355.6)
Id effect	YES	YES	YES	YES
Year effect	YES	YES	YES	YES
N	126	153	90	189
R^2	0.477	0.265	0.729	0.281

Note: The numbers in the brackets are standard errors; ***, ** and * indicate a statistical significance at the 1%, 5%, and 10% levels, respectively. "No" means that the effect was not controlled; "Yes" means that the effect was controlled.

4.4.2. Differentiate Carbon Emission Intensity

Due to the vastness of China and the significant differences in the economic development and environmental conditions across its provinces, the use of industrial robots may have varying effects on the technical complexity of the energy sector exports across the different regions. In light of this, we referred to Shi and Liu (2022) [38] for a regional division based on the carbon emission intensity of each province [39]. Specifically, the Class I region, which comprises ten provinces, including Ningxia, Inner Mongolia, Xinjiang, Guizhou, Shanxi, Hebei, Qinghai, Gansu, Jilin, and Liaoning, is characterized by a high carbon intensity. On the other hand, the Class II region, which includes 21 provinces such as Heilongjiang, Shaanxi, Anhui, Yunnan, Guangxi, Shandong, Henan, Tianjin, Hubei, Jiangxi, Chongqing, Hunan, Hainan, Sichuan, Jiangsu, Fujian, Zhejiang, Guangdong, Shanghai, Tibet, and Beijing, has a relatively lower carbon intensity level.

Following the regional division results, we conducted additional grouping tests to explore the issue of regional heterogeneity in the impact of industrial robots on the technical complexity of the energy exports. Table 6 presents the regression results of the industrial robots' impact on the technical complexity of the energy exports in the Class I and Class II regions. Columns (3) and (4) of Table 6 indicate that the regression coefficient of the industrial robots' application on the technical complexity of the exports was significantly positive in the Class II regions with a low to medium carbon intensity. This suggests that the application of industrial robots significantly enhanced the level of the international division of labor in energy in the Class II regions and increasing the level of the industrial robot application can effectively improve the technical complexity of the energy exports in the region. The reason for this may be that, compared to the Class I regions, the Class II regions have a more reasonable industrial structure, a relatively higher level of production technology, a more reasonable distribution of high-tech and modern service industries, and a more efficient and reasonable application of industrial robots. It is worth noting that the impact of industrial robots on the technical complexity of the exports in the Class I regions was not significant. This may be due to the higher carbon intensity and less advanced industrial structure of these regions, resulting in a less efficient and less effective application of industrial robots in these regions.

5. Mechanism Inspection

The empirical findings of this study demonstrated that the adoption of industrial robots contributed to an increase in the technical complexity of the exports, and this relationship was robust. The second part of the theoretical mechanism analysis posited that the use of robots could have a "cost-saving effect" on a firm's export sophistication. Specifically, our results indicated that the use of robots could impact a firm's variable costs by enhancing its productivity or affecting its fixed production costs by reducing overhead expenses, ultimately influencing the firm's export technological complexity. Therefore, this study examined the mechanism from both the efficiency and cost perspectives.

To test the theoretical Hypothesis 2, we introduced the following model.

$$Element_{it} = \alpha_0 + \alpha_1 Robots_{it} + \alpha_2 X_{it} + \mu_i + \gamma_t + \varepsilon_{it} \quad (6)$$

$$Fixcost_{it} = \alpha_0 + \alpha_1 Robots_{it} + \alpha_2 X_{it} + \mu_i + \gamma_t + \varepsilon_{it} \quad (7)$$

where $Element_{it}$ represents the element structure, $Fixcost_{it}$ represents the export cost, and the other control variables are consistent with the benchmark regression above.

5.1. Element Structure

According to the economic literature, the factor structure of an industry reflects its endowment status and comparative advantage in production. In particular, if an industry has a significant share of capital in its factor structure, then it must improve the quality of its products and increase its exports based on its endowment advantage in order to achieve an international division of labor status. This proposition is supported by the previous research [23,40]. Therefore, it is crucial for industries to recognize their factor endowments and leverage them effectively in order to compete successfully in the global market.

The adoption of industrial robots can be viewed as an increase in capital inputs, which may be more prevalent in industries with a higher share of capital relative to those with a higher share of labor. This can obscure the impact of the robots on quality improvement through a greater share of the capital inputs. However, the factor structure of an industry depends not only on the total amount of capital inputs but also on the relative proportions of the capital and labor inputs. Thus, the changes in the labor inputs are equally important. Although robots affect the factor structure through the capital factor, the impact on labor should not be overlooked.

The application of industrial robots affects the complexity of the export technology for two main reasons. First, in terms of the capital factor, as the scale and variety of robot applications expand and the level of robot use in the production process of enterprises continues to increase, capital-intensive production tasks become more complex. This factor-biasing pattern leads enterprises to favor the use of capital to produce high-quality products, which ultimately results in an increase in the technological complexity [23,40]. Second, in terms of the labor factors, the "substitution" and "creation" theories determine the impact of the robots on labor. The "substitution theory" posits that robots are designed for specific production needs and can perform relatively fixed tasks, so an increase in the number of robots will inevitably displace human labor and result in the "man-for-machine" phenomenon. In contrast, the "creation theory" suggests that the application of robots can expand the scale of production and increase consumer demand, leading to an expansion of production and an increase in the demand for labor. Additionally, a large-scale application of robots requires the support of related skills, which can create new job opportunities.

Therefore, the application of industrial robots can alter the structural elements of an industry and affect the technological complexity of enterprises. This paper defines the factor structure (*Element*) as the ratio of fixed assets to the number of employees employed, drawing from Yuan et al. (2022) [41].

5.2. Reducing Export Costs

Although the introduction of robots can initially increase costs and exacerbate the financing constraints for firms, the advantages of the low-variable costs associated with their usage gradually emerge over time. Specifically, industrial robots can replace low-skilled labor and reduce wage expenses for firms, while also squeezing the market space for low-skilled labor, which indirectly reduces the average wage level and production costs. These benefits are especially significant for export-oriented firms, where the demand for low-skilled labor, such as handling and warehousing, is urgent. Consequently, the application of industrial robots can lower export costs and free up capital for technology research and development, which in turn upgrades the enterprise's export products.

However, when measuring export costs, there may be deviations between the actual costs and the book records of enterprises, as the standards for measuring various cost items incurred during the export process are not always uniform. To address this, this study followed Fu and Lu (2021) [42] and used a proxy variable for the fixed production costs (*Fixcost*) as the sum of the financial, administrative, and selling costs.

In Table 7, columns (1) and (2) examine the impact of the robot usage on the factor structure and fixed costs, respectively. The coefficient of the core explanatory variable was significantly positive in column (1), indicating that the application of industrial robots optimized the enterprise factor structure, promoted capital deepening, and adjusted the labor structure. Meanwhile, the regression coefficient in column (2) was also significantly positive, suggesting that the use of industrial robots could optimize the export costs. Therefore, it can be inferred that the main mechanisms driving the use of robots in industry to improve the technical complexity of the exports are the factor structure optimization and fixed cost reduction, which confirms Hypothesis 2.

Table 7. Test results of the action mechanism.

	(1)	(2)
	Element	*Fixcost*
Robots	0.0148 ** (0.00593)	0.0117 *** (0.00394)
Develop	3.037 (9.638)	−5.034 (6.395)
FDI	0.00507 (0.00509)	0.00247 (0.00338)
Patent	−0.000946 *** (0.000358)	−0.000186 (0.000237)

Table 7. *Cont.*

	(1)	(2)
	Element	*Fixcost*
Labor	−0.0245 (0.0267)	0.0273 (0.0177)
Index	2.895 *** (0.655)	−0.367 (0.435)
Fin	−0.0954 *** (0.0246)	−0.0788 *** (0.0163)
_cons	−162.0 (1019.6)	757.5 (676.4)
Id	YES	YES
Year	YES	YES
N	279	279
R^2	0.567	0.238

Note: The numbers in the brackets are standard errors; ***, ** indicate a statistical significance at the 1%, 5% levels, respectively. "No" means that the effect was not controlled; "Yes" means that the effect was controlled.

6. Conclusions and Policy Implications

The utilization of industrial robots undoubtedly plays a significant role in enhancing the Chinese energy industry's level of participation in the international division of labor. This paper examined the empirical impact of industrial robots on the level of participation and the underlying mechanisms from the perspective of their technological complexity. To accomplish this objective, we employed the data on the robots from various countries provided by the IFR and several databases such as the EPS. Our research findings indicated that industrial robotics enhances the level of participation in the international division of labor in China's energy industry. We further observed a regional heterogeneity whereby the impact of industrial robotics on the export technology complexity was more pronounced in the regions with an abundant energy production and a low carbon emission intensity. Moreover, the promotion effect was most significant for the regions with an abundant energy production. Lastly, the application of industrial robotics improved the technical complexity of the exports by optimizing the factor structure and reducing the export costs. In light of the widespread adoption of robotics, this paper established a theoretical link between the robotics application and the energy industry's participation level in the international division of labor. Furthermore, we provided evidence based on the perspective of the export technology complexity, which offered valuable policy insights to enhance the Chinese energy industry's high-quality development and enable Chinese energy enterprises to realize their status in the international division of labor system.

Based on the findings of this study, the following recommendations are proposed.

Firstly, in order to enhance the participation of the energy industry in the international division of labor, the cost of applying industrial robotics technology needs to be reduced to enable its large-scale application in the sector. Although industrial robots have been widely used in the energy industry, their high price and lack of scale have impeded their deep integration with the sector. Therefore, it is essential to further reduce the cost of applying industrial robot technology and promote its integration with the energy industry to meet the need for large-scale production operations. This can be achieved by optimizing and upgrading the entire production process, from R&D to design, processing, and sales, and by enhancing the use of emerging technologies in economically underdeveloped regions to transform the traditional energy industry into an intelligent and digitized system, thereby improving the competitiveness of the energy products in the global market.

Secondly, more effective and transparent theoretical policies should be formulated to improve the international division of labor in the energy industry. To ensure the long-term stability of an enterprise's investment in the research and development of related technologies, subsidies for the purchase of industrial robots and other related technologies should be increased to reduce the financial pressure on the enterprises during the preliminary research and development stage. Additionally, education in the disciplines related to industrial robots should be deepened, and more highly skilled talents should be cultivated to match the intelligent development of the enterprises. Finally, the government can play a leading role in coordinating and advocating for the commercial platform of industrial robotics, improving the transparency of information for both R&D and the use of robots. Additionally, the government can encourage financial institutions to invest in SMEs to reduce the mismatch of funds, thereby enabling more SMEs to enhance their R&D and innovation capabilities through advanced technologies, such as industrial robots, and to achieve a higher level of participation in the international division of labor for Chinese energy enterprises.

It is important to note that the energy data used in this study were mainly derived from the customs and industrial enterprise databases, and the export technical complexity of the energy industry was measured by dividing the energy into six categories. However, due to the limitation of a statistical caliber, the current measurement of the export technical complexity of the regional energy industries was limited. To address this limitation, future research should focus on narrowing the statistical caliber and accessing the data related to the robots of non-listed enterprises, thus enabling data matching between the number of industrial robots and energy enterprises, resulting in a more detailed analysis.

Author Contributions: Conceptualization, J.P.; methodology, W.Z.; software, W.Z.; validation, J.P., W.Z. and L.Z.; formal analysis, J.P. and W.Z.; investigation, L.Z.; resources, J.P.; data curation, W.Z.; writing—original draft preparation, W.Z.; writing—review and editing, J.P., W.Z. and L.Z.; supervision, J.P.; funding acquisition, J.P. and W.Z. All authors have read and agreed to the published version of the manuscript.

Funding: This research was funded by the MOE (Ministry of Education in China) Project of Humanities and Social Sciences and the Soft Science Research Project (NO. 18YJCZH029). The authors would also like to acknowledge the financial support of the Youth Talent Project from the Wuhan-Shuguang Project (No. 2022010801020365), the Research Fund Project from the Wuhan Institute of Technology (NO. K202248), the Statistical Research Project from the National Bureau of Statistics in China (NO. 2022LY057), the Social Science Foundation of Hubei Province (NO. HBSK2022YB339, HBSK2022YB336), the Social Science Foundation from the Wuhan Institute of Technology (NO. R202105), and the Graduate Education Innovation Fund Project from the Wuhan Institute of Technology (CX2022388).

Data Availability Statement: The data presented in this study are available on request from the corresponding authors.

Acknowledgments: We are grateful for the Energies Office and the comments and suggestions from the editor and anonymous reviewers who helped improve the paper. We would also like to express our gratitude to the Ministry of Education in China, the Knowledge Innovation Program of the Wuhan Shuguang Project, and the Research Fund Project of the Wuhan Institute of Technology for providing funding support.

Conflicts of Interest: The authors declare no conflict of interest.

References

1. Jiang, F. Contributing Wisdom and Strength to the Progress of Human Civilization: The Development Concept and Great Practiceof the CPC's Maintenance of a Global Vision. *J. Manag. World* **2022**, *38*, 41–49.
2. Zhou, S.; Hu, R. Research on Stochastic Comvergence of High Quality Energy Development in China. *J. Hunan Univ.* **2022**, *36*, 66–76.
3. Geng, Y.; Bai, L. Human Capital Structure Upgrading, R&D Intensity and Upgrading in Global Value Chain of Manufacturing Industry. *World Econ. Stud.* **2019**, *8*, 88–102+136.
4. Yang, B.; Liu, B.; Peng, J.; Liu, X. The impact of the embedded global value chain position on energy-biased technology progress: Evidence from chinas manufacturing. *Technol. Soc.* **2022**, *71*, 102065. [CrossRef]
5. Acemoglu, D.; Lelarge, C.; Restrepo, P. Competing with Robots: Firm-Level Evidence from France. *AEA Pap. Proc.* **2020**, *110*, 383–388. [CrossRef]

6. Acemoglu, D.; Restrepo, P. Automation and New Tasks: How Technology Displaces and Reinstates Labor. *J. Econ. Perspect.* **2019**, *33*, 3–30. [CrossRef]
7. Cai, Z.; Qi, J. Does the Adoption of Industrial Robots Upgrade the Export Product Quality—Evidence from Chinese Manufacturing Enterprises. *J. Int. Trade* **2021**, *10*, 17–33.
8. Graetz, G.; Michaels, G. Robots at Work. *Rev. Econ. Stat.* **2018**, *100*, 753–768. [CrossRef]
9. Acemoglu, D.; Restrepo, P. Robots and jobs: Evidence from US labor markets. *J. Polit. Econ.* **2020**, *128*, 2188–2244. [CrossRef]
10. Peng, J.; Fu, S.; Gao, D.; Tian, J. Greening China's Growth: Assessing the Synergistic Impact of Financial Development and Technological Innovation on Environmental Pollution Reduction—A Spatial STIRPAT Analysis. *Int. J. Environ. Res. Public Health* **2023**, *20*, 5120. [CrossRef]
11. Chen, Y.; Lin, C.; Chen, X. Artificial Intelligence, Aging and Economic Growth. *Econ. Res. J.* **2019**, *54*, 47–63.
12. Lin, C.; Chen, X.; Chen, W.; Chen, Y. Artificial Intelligence, Economic Growth and Household Consumption Improvement: A Capital Structure Perspective. *China Ind. Econ.* **2020**, *2*, 61–83.
13. Peng, J.; Chen, H.; Jia, L.; Fu, S.; Tian, J. Impact of Digital Industrialization on the Energy Industry Supply Chain: Evidence from the Natural Gas Industry in China. *Energies* **2023**, *16*, 1564. [CrossRef]
14. Goldfarb, A.; Trefler, D. *AI and International Trade*; National Bureau of Economic Research: Cambridge, MA, USA, 2018.
15. Artuc, E.; Bastos, P.; Rijkers, B. *Robots, Tasks, and Trade*; CEPR Discussion Paper No. DP14487; World Bank: Washington, DC, USA, 2020.
16. Trajtenberg, M. *AI as the Next GPT: A Political-Economy Perspective*; National Bureau of Economic Research: Cambridge, MA, USA, 2018.
17. Tian, Y.; Zhuo, Y.; Zou, H.; Wwang, L. The Impact of Artificial Intelligence Technological Revolution on International Trade. *Intertrade* **2020**, *2*, 24–31.
18. Lu, H.; Peng, J.; Lu, X. Do Factor Market Distortions and Carbon Dioxide Emissions Distort Energy Industry Chain Technical Efficiency? A Heterogeneous Stochastic Frontier Analysis. *Energies* **2022**, *15*, 6154. [CrossRef]
19. Lu, W.; Meng, X. The Application of Industrial Robots, the Adjustment of Employment Market Structure and the Development of Service Trade. *Int. Econ. Trade Res.* **2021**, *37*, 4–20.
20. Liu, P.; Xia, Y. Do the Pilot Free Trade Zones Increase Regional Export Techology Complexity? *J. Cap. Univ. Econ. Bus.* **2022**, *24*, 42–51.
21. Liu, Z.; Lin, H.; Zhang, S. Urban Digital Technology, Innovation Heterogeneity and Export Product Technical Complexity of "Hidden Champion" Enterprises. *Contemp. Financ. Econ.* **2021**, *10*, 103–116.
22. Cockburn, I.; Henderson, R.; Stern, S. *The Impact of Artificial Intelligence on Innovation*; National Bureau of Economic Research Inc.: Cambridge, MA, USA, 2018.
23. Li, Y.; Lu, Y.; Wang, T. How do Financial Support and Technological Innovation Affect Export Sophistication? An Empirical Research Based on China's High-Tech Industry. *Foreign Econ. Manag.* **2019**, *41*, 43–57.
24. Graetz, G.; Michaels, G. Robots at work: The impact on productivity and jobs. *The Magazine for Economic Performance*, 18 March 2015.
25. Hausmann, R.; Hwang, J.; Rodrik, D. What you export matters. *J. Econ. Growth* **2007**, *12*, 1–25. [CrossRef]
26. Zhang, A.; Yin, M. Technological Innovation, Demographic Structure and Export Sophistication of China's Manufacturing Industries. *Soft Sci.* **2019**, *33*, 29–34.
27. Hicks, J.R. Economic Foundations of Wage Policy. *Econ. J.* **1955**, *65*, 389–404. [CrossRef]
28. Li, J.; Zhao, X. Accelerated Depreciation Policy of Fixed Assets and Capital-Labor Ratio. *Financ. Trade Econ.* **2021**, *42*, 67–82.
29. Xu, B.; Lu, J. Foreign direct investment, processing trade, and the sophistication of China's exports. *China Econ. Rev.* **2009**, *20*, 425–439. [CrossRef]
30. Wang, Y.; Dong, W. How the Rise of Robots Has Affected China's Labor Market: Evidence from China's Listed Manufacturing Firms. *Econ. Res. J.* **2020**, *55*, 159–175.
31. Xu, Y.; Zhu, J.; Tan, C. Intelligent Manufacturing, Labor Skill Structure an Export Technology Complexity. *Financ. Trade Res.* **2022**, *33*, 16–27.
32. Ma, X.; Liu, Y.; Li, Z. Research on the Countercurrent of Talents in China from the Perspective of "Safe House of Talents". *Sci. Technol. Manag. Res.* **2019**, *39*, 141–148.
33. Wen, Y.; Xiao, J.; Peng, J. The effects of the "Zero Routine Flaring by 2030" initiative: International comparisons based on generalized synthetic control method. *Environ. Impact Asses.* **2023**, *100*, 107095. [CrossRef]
34. Hong, Y.; Zhou, Y. Overseas BachGround of Senior Management Team, Financial Mismatch and Total Factor Productivity of Enterprises. *J. Harbin Univ. Commer.* **2022**, *4*, 93–104+118.
35. Hu, X.; Shi, B.; Yang, J. Mechanism Identification and Empirical Evidence of Digital Economy Strengthening the Development of Real Economy. *Econ. Probl.* **2022**, *12*, 2414.
36. Liu, X.; Zou, K. Does Sci-tech Finance Promote the Synergy of Economy and Innovation? *J. Hunan Univ. Sci. Technol.* **2021**, *24*, 71–81.
37. Yan, X.; Li, W.; Gao, R. Impact of Artificial Intelligence on China's Labor Market. *Rev. Ind. Econ.* **2021**, *2*, 65–77.
38. Shi, M.; Liu, N. Can Taxation Greening Reduce Carbon Emissions?—An Empirical Analysis Based on Provincial Panel Data in China. *Fisc. Sci.* **2022**, *10*, 98–112.

39. Peng, J.; Wen, L.; Mu, X.; Xiao, J. The evolving centres of gravity in China's oil and gas industry: Evidence from infrared radiation imaging gas flaring data. *Energy Sustain. Dev.* **2023**, *73*, 263–279. [CrossRef]
40. Hallak, J.C.; Sivadasan, J. Product and process productivity: Implications for quality choice and conditional exporter premia. *J. Int. Econ.* **2013**, *91*, 53–67. [CrossRef]
41. Yuan, Q.; Ji, Y.; Yu, S. Does artificial intelligence help manufacturing companies upgrade their exports? Analysis based on technology complexity perspective. *Rev. Ind. Econ.* **2022**, *3*, 69–82.
42. Fu, D.; Lu, C. Service Liberalization Promotes Transformation of Trade Modes—Firm-level Theory and Evidence from China. *China Ind. Econ.* **2021**, *7*, 156–174.

energies

Article

The Impact of Export Sophistication of the New Energy Industry on Carbon Emissions: An Empirical Study

Ke Huang [1], Teng Wang [1,*], Jiachao Peng [2,3] and Lijun Sun [1]

1 School of Economics and Management, China University of Geosciences, Wuhan 430074, China
2 Law and Business School, Wuhan Institute of Technology, Wuhan 430205, China
3 Center for High Quality Collaborative Development of Resources, Environment and Economy, Wuhan Institute of Technology, Wuhan 430205, China
* Correspondence: wangteng@cug.edu.cn

Abstract: Existing research has insufficiently explored the nexus between the new energy industry and CO_2 emissions from the standpoint of export sophistication. This study analyses the implications of the new energy industry's export sophistication on CO_2 emissions, regional heterogeneity, and its influencing mechanism by gathering data from 31 major economies throughout the world between 1996 and 2021. The study found that the new energy industry's export sophistication helps reduce carbon dioxide emissions, and this conclusion still holds after robustness testing; the carbon emission reduction effect of the export sophistication of the new energy industry is more significant in developed countries than in developing countries; the new energy industry's export sophistication possesses a crowding-out effect on domestic technological progress, which to a certain extent impedes carbon reduction effect. This paper's findings provide theoretical guidance for the global low-carbon energy transition.

Keywords: carbon emissions; export sophistication; new energy industry; influential mechanism; heterogeneity; fixed effects model; mediation effect model

Citation: Huang, K.; Wang, T.; Peng, J.; Sun, L. The Impact of Export Sophistication of the New Energy Industry on Carbon Emissions: An Empirical Study. *Energies* **2023**, *16*, 3846. https://doi.org/10.3390/en16093846

Academic Editor: Wen-Hsien Tsai

Received: 31 January 2023
Revised: 21 April 2023
Accepted: 27 April 2023
Published: 29 April 2023

1. Introduction

The unprecedented globalization of international energy commerce in the past few decades has significantly contributed to the growth and prosperity of the global economy. Unfortunately, the fossil fuel-based energy trade structure has also emitted a large quantity of carbon dioxide (CO_2), resulting in global warming, which has posed a grave danger to human survival and development [1]. New energy, also known as unconventional energy, refers to non-traditional forms of energy, including solar, wind, biomass, geothermal, hydroelectric, and nuclear energy. Compared to traditional energy, new energy has the advantages of pure environmental protection, abundant reserves, and sustainability, which are crucial for resolving the severe environmental pollution problems and the greenhouse effect in the world today [2,3]. Statistics from *China's National Energy Administration* show that China's power production from renewable energy in 2022 is equivalent to lowering domestic CO_2 by approximately 2.26 billion tons and exporting wind power photovoltaic products to decrease CO_2 for other countries by nearly 573 million tons for a total reduction of 2.83 billion tons [4].

Despite worldwide governmental recognition of the potential for new energy to reduce carbon emissions, the latest data from BP's 2022 World Energy Statistics Review indicates that the global energy trade continues to be dominated by fossil fuels, including coal, oil, and natural gas, with new energy exports receiving notably less emphasis. This is mainly because new energy has a higher use cost than traditional fossil energy, and its export is heavily affected by policies, which makes it less competitive [5]. Due to the limited number of new energy exports, researching and enhancing the export sophistication of new energy, which demonstrates how competitive new energy is, is an additional effective strategy for attaining global carbon reduction goals [6].

Literature abounds with studies investigating the connection between CO_2 and new energy. The prevalent theory in academia is that increased energy use may adversely decrease carbon emissions [7–9]. Dong et al. (2018) [10] investigated the link between the new energy industry development and CO_2 and found that new energy development may considerably lower carbon dioxide emissions. The findings of Acheampong et al. (2022) [11], Habiba et al. (2022) [12], Rahman and Alam (2022) [13], Djellouli et al. (2022) [14] corroborate the conclusion that the new energy may contribute to the carbon reduction. In contrast to the conclusion that new energy can help reduce carbon dioxide emissions, Al-Mulali et al. (2015) [15] found that Vietnam's use of renewable energy has an insignificant impact on decreasing carbon dioxide emissions, and Zaidi et al. (2018) [16] came to the same conclusion in their sample of Pakistan. Additionally, Jebli and Youssef (2017) [17] found that long-term carbon dioxide emissions in the five nations of North Africa had grown due to the use of renewable energy.

Existing research on new energy and CO_2 primarily examines the impact of new energy on CO_2 from the perspective of total new energy use, while few scholars investigate its carbon reduction effect from the perspective of new energy competitiveness. Moreover, the contradictory conclusion between new energy and CO_2 indicates that more in-depth research on the relationship is required. Based on the existing literature, this study investigates the relationship between export sophistication of new energy and carbon dioxide, investigates the influence mechanism between the two, and examines whether this relationship exhibits regional heterogeneity.

This paper's contribution to the existing body of literature is summarized in three points. As an important indicator of new energy competitiveness, this study evaluates the new energy industry's export sophistication in 31 significant economies from 1996 to 2021 and empirically tests whether there is a carbon emission reduction effect using a fixed-effect model. Second, in order to avoid the similar phenomenon of the mixed conclusion of new energy and CO_2, we employ the mediation effect model to analyze in depth the mechanism of new energy export sophistication on carbon emissions, which has significant theoretical significance in terms of revealing the black box between them. Thirdly, there are numerous differences between countries, including economic development, the potential for new energy development, etc. Therefore, it is more plausible to analyze the regional heterogeneity of carbon emission reduction in the export sophistication of new energy, and this is useful for making emission reduction recommendations.

The remainder of the article is divided into six sections. Section 2 organizes the extant literature on the export sophistication of new energy and carbon dioxide. In Section 3, variable selection, data sources, and model methodology are introduced. Sections 4 and 5 illustrate the findings, mechanism, and regional heterogeneity of the impact of the export sophistication of new energy on carbon emissions. Section 6 contains the research findings and proposed countermeasures.

2. Literature Review

2.1. *Research into the Export Sophistication of New Energy Industry*

Export sophistication is a critical indicator for measuring the structure of national or regional export commodities, as it reflects the competitive advantage of export commodities. The introduction of export sophistication can be traced back to Michaely's (1984) [18] trade specialization index. The indicator implies that the degree of technology incorporated in an exported commodity is proportional to the per capita income of the exporting country. Hausmann et al. (2007) [19] took the initiative in elucidating the connotation of export sophistication and employing it as a measurement of the structure of export products. The greater the indicator value, the greater the likelihood that the product can achieve a competitive advantage in the face of fierce market rivalry. Since its proposal, export sophistication has shifted the emphasis of international commodity trade competition from export quantity to export competitiveness. With the expansion of research on export sophistication, different levels of export sophistication have been implemented. At the national level, Jarreau and Poncet (2012) [20] computed the export sophistication of 30 provinces in China from 1997 to 2009; Rehman et al. (2023) [21] assessed the export sophistication of renewable and non-renewable energy in OECD countries during 1990-2019, respectively; At the industrial level, Su et al. (2020) [22] took the manufacturing industry as the research object, calculating the sophistication of manufacturing exports in 36 countries from 2005 to 2014. At the enterprise level, Song et al. (2022) [23] assessed the export sophistication of 498,945 Chinese manufacturing enterprises by combining the Chinese customs database with the Chinese industrial enterprise database.

Existing research has provided a comprehensive discussion of the definition and measurement of export sophistication, and research on export sophistication involves different groups, including the nation, industry, and enterprise levels. However, research on export sophistication in the new energy industry is scarce. Zheng and Wang (2019) [24] used the United Nations Comrade database to measure the new energy industry's export sophistication in 30 countries around the world from 2000 to 2015, comparing and analyzing the evolution of the export sophistication of transnational new energy industries and their subdivisions. Cao et al. (2019) [25] calculated the dynamic changes in the export sophistication of China's new energy industry from 2007 to 2016 and discovered that the overall new energy industry's export sophistication exhibited a fluctuating growth trend, the proportion of high-tech sophistication was low, and the export structure exhibited a deteriorating trend.

2.2. Studies of Carbon Dioxide

The methods for calculating carbon dioxide emissions concentrate primarily on three factors: the measuring method, the material balance method, and the carbon emission factor method. The measuring method uses the velocity, concentration, and flow rate of carbon dioxide sample emissions to calculate the total quantity of carbon emissions [26]. This method necessitates sophisticated measuring instruments and is primarily employed by environmental monitoring departments. Material balance is an approach for calculating the total quantity of carbon dioxide emissions based on the input and output material conservation theorem. This method requires maximum control over the enterprise's production and emission situation [27]. Based on the *2006 IPCC National Greenhouse Gas Inventory Guidelines* issued by the Intergovernmental Panel on Climate Change (IPCC), the carbon emission factor method multiplies and accumulates various forms of energy consumption and their respective carbon emission factors to obtain carbon emissions. This method is considered the most authoritative carbon emission accounting method in the world [28]. It serves as a significant foundation for countries to report their carbon emissions to the IPCC.

Scholars have been interested in the influencing variables of carbon dioxide emissions for a very long time. The relationship between economic growth and carbon emissions is one of the contemporary research hotspots, and the environmental Kuznets hypothesis is the main focus of related research. According to Ridzuan et al. (2020) [29], Malaysia's long-term economic growth and carbon emissions show an inverted U-shape. As the economy expands, carbon dioxide emissions first rise before starting to fall once they reach a critical threshold. The effect of urbanization on carbon emissions has received significant attention in terms of population growth. Sufyanullah et al. (2022) [30] discovered that the progress of urbanization in Pakistan has resulted in a rise in carbon dioxide emissions. The conclusion that there is a positive association between urbanization level and CO_2 also pertains to the Philippines [31]. One of the key elements impacting carbon emissions is foreign direct investment. According to Lu et al. (2023) [32], there is a pollution refuge in transition economies since there is a positive association between foreign direct investment and carbon emissions. The literature on export sophistication and carbon emissions is abundant, whereas the literature on examining carbon emissions from the perspective of export sophistication in the new energy industry is extremely scarce. Based on previous research, we investigate the relationship between the new energy industry's export sophistication and carbon dioxide, as well as the impact Mechanism of export sophistication on carbon emissions and potential heterogeneity in carbon emission reduction.

3. Methods

3.1. Model Specification

3.1.1. Econometric Model

This study employed a model with fixed effects [23] to investigate the impact of the new energy industry's export sophistication on CO_2. The econometric model is shown in Equation (1).

$$\ln CO_{2_{it}} = \alpha + \beta_1 \ln EXPY_{it} + \beta_2 \ln FDI_{it} + \beta_3 \ln IT_{it} + \beta_4 \ln Urb_{it} + \delta_i + \varphi_t + \varepsilon_{it} \quad (1)$$

where $\ln CO_2$ and $\ln EXPY$ are the logarithms of Carbon Dioxide and export sophistication; $\ln FDI$, $\ln IT$ and $\ln Urb$ are the logarithms of control variables, namely foreign direct investment, international trade and urbanization; i and t represent the country and year respectively; α is a constant term; β_1 and β_2-β_4 are the coefficients of $\ln EXPY$ and 3 control variables $\ln FDI$, $\ln IT$, and $\ln Urb$ on $\ln CO_2$, respectively. δ_i and φ_t represent national fixed effects and temporal fixed effects, respectively; ε_{it} represents the random error term.

3.1.2. Mediation Effect Model

Furthermore, we use the mediation effect model to find out how the export sophistication of new energy exports impacts carbon emissions [33]. The 3-step regression technique is suggested to assess if technological progress has a mediating influence with the aid of Baron and Kenny (1986) [34].

$$\ln CO_{2_{it}} = \alpha_0 + \alpha_1 \ln EXPY_{it} + \alpha_2 \ln X_{it} + \delta_i + \varphi_t + \varepsilon_{it} \quad (2)$$

$$\ln TP_{it} = \phi_0 + \phi_1 \ln EXPY_{it} + \phi_2 \ln X_{it} + \delta_i + \varphi_t + \nu_{it} \quad (3)$$

$$\ln CO_{2_{it}} = \gamma_0 + \gamma_1 \ln EXPY_{it} + \gamma_2 \ln TL_{it} + \gamma_3 \ln X_{it} + \delta_i + \varphi_t + \tau_{it} \quad (4)$$

where $\ln TP$ is the logarithm of technological progress; α_1 in Equation (2) is the total effect of the $\ln EXPY$ on the $\ln CO_2$; ϕ_1 in Equation (3) is the effect of $\ln EXPY$ on the $\ln TP$; In Equation (4), the coefficient γ_1 is the direct effect of the $\ln EXPY$ on the $\ln CO_2$ after controlling for the influence of the $\ln TP$. X_{it} is the control variable mentioned above; ε_{it}, ν_{it}, and τ_{it} are random error terms.

The intermediary effect of the explanatory variable $\ln EXPY$ on the $\ln CO_2$ is $\phi_1 \times \gamma_2$, and the relationship between the total effect, the intermediary effect, and the direct effect is:

$$\alpha_1 = \gamma_1 + \phi_1 \times \gamma_2 \quad (5)$$

3.2. Variables and Data

3.2.1. Explained Variable

The explained variable in this study is CO_2, and it has two measurement indicators: total carbon dioxide emissions [35,36] and per capita carbon dioxide emissions [37]. The former is an absolute number, while the latter is a relative one. We ultimately settled on the total carbon emissions as the indicator to measure CO_2 and used the per capita carbon dioxide emissions for the robustness test. This was done because the carbon emission reduction and carbon neutralization policies developed by nations around the world are based on the actual situation of total carbon dioxide emissions. The indicator value that is lower indicates lower national carbon dioxide emissions, and vice versa. The *2022 BP Statistical Review of World Energy* (accessed on 6 September 2022, from https://www.bp.com/en/global/corporate/energy-economics/statistical-review-of-world-energy/co2-emissions.html) provides information on total carbon emissions.

3.2.2. Explanatory Variable

Export sophistication depicts the degree of productivity connected with a national or regional array of exported commodities [38]. We rely on Hausmann et al.'s (2007) [19] measuring approach to calculate export sophistication.

First, assuming the export sophistication of the exported good k is $PRODY_k$, which is calculated as follows:

$$PRODY_k = \sum_{k \in K_i} \frac{x_{ik}/X_i}{\sum_{j \in K_i} x_{ik}/X_i} \times Y_i \tag{6}$$

where i refers to the country or region i that exports goods k; x_{ik} represents the total export value of goods k in country i; X_i refers to the total export value of country i; and x_{ik}/X_i represents the proportion of goods k export value in the total export of that country; Y_i is the per capita GDP of country or region i; and k_i denotes the collection of all countries exported good k; $PRODY_k$ is the sum of the product of the export proportion of each country's good k and the country's PGDP.

Based on the calculation of $PRODY_k$, we further assess the export sophistication of industry j. Considering that N represents the total number of exported goods k produced by industry j of country i, the export sophistication level $EXPY_{ji}$ of industry j in country i is as follows:

$$EXPY_{ji} = \sum_{k \in N} s_{kji} \times PRODY_k \tag{7}$$

Among them, s_{kji} is the share of the export value of good k in the total export value of industry j in country i. $EXPY_{ji}$ is essentially the weighted average sum of $PRODY_k$ in industry j of country i.

We used Wang et al.'s (2017) [5] assessment and categorization of the new energy industry to calculate its export sophistication. The new energy industry is comprised of four subindustries: wind energy, solar energy, biomass energy, and nuclear power technology. The HS1996 standard is adopted by the appropriate goods and classification codes. The original data on export value related to the HS 6-bit code in these four industries are all taken from the *UN Comtrade database* (accessed on 29 August 2022, from https://comtrade.un.org/data/). The raw data on total exports of various countries and PGDP are taken from the *World Bank Open Data* (accessed on 1 September 2022, from https://data.worldbank.org.cn/indicator).

3.2.3. Control Variables

(1) Foreign direct investment (FDI)

As 1 of the most influential variables on carbon emissions, the influence of FDI on CO_2 has been the subject of extensive academic study. The pollution refuge theory and the pollution halo hypothesis are 2 competing theories about how FDI affects CO_2 emissions. According to the pollution haven hypothesis, developing countries tend to adopt lower environmental protection standards in order to attract more FDI, which brings in a lot of low-quality, pollution-intensive FDI and turns developing countries into the sources of developed countries' carbon emissions [39,40]; whereas the pollution halo hypothesis contends that FDI brings advanced technology and a wealth of management experience to host nations [41,42]. According to Ali et al. (2023) [43], we utilized the net inflow of FDI as a gauge of a country's ability to attract FDI. The original data regarding net FDI inflows were obtained from *World Bank Open Data* (accessed on 1 September 2022 at https://data.worldbank.org.cn/indicator).

(2) International Trade (IT)

The global trade system has altered as a result of increasing global economic integration, which has also sparked studies on how trade affects carbon emissions. The impact of international trade on CO_2 is currently primarily focused on 2 aspects: on the 1 hand, international trade encourages global economic growth through the specialized division of labor, which increases energy consumption and, in turn, increases total carbon dioxide emissions [44]; on the other hand, international trade enhances international exchange and cooperation, particularly promoting technology exchange between different countries, which helps to reduce global carbon emissions [45]. The *World Bank Open Data* is utilized to extract the pertinent statistics, which are used to measure total goods import and export trade (accessed on 1 September 2022, from https://data.worldbank.org.cn/indicator).

(3) Urbanization (Urb)

Another important element that has an impact on carbon emissions is urbanization (Urb). On the 1 hand, both centralized energy use and information spillover, as well as technological advancements brought about by urbanization, contribute to improving energy use efficiency and lowering CO_2 [46]. On the other hand, the advancement of urbanization will improve urban population density and the resulting increase in urban infrastructure, which leads to an increase in CO_2 to some extent [30,47]. According to Wang et al. (2021) [48], the ratio of the urban population to the overall population is chosen to properly depict the degree of urbanization. The raw data involved in the control variables are derived from the *World Bank Open Data* (accessed on 1 September 2022, from https://data.worldbank.org.cn/indicator).

3.2.4. Intermediate Variable

As 1 of the major determinants of a country's carbon dioxide emissions, the technological progress (TP) of the host nation has been the subject of extensive study. Technological progress at the source of energy consumption in host countries can reduce CO_2 production by substituting fossil fuels with clean energy; technological progress during the consumption of energy can reduce CO_2 production by increasing energy efficiency; and technological progress at the end of pollution emissions can revert CO_2 emissions through carbon capture and storage. Overall, technological progress in host nations contributes to the reduction of carbon emissions [49]. We determine the host country's overall technical advancement using the Cobb-Douglas production function [50]. The *World Bank Open Data* (accessed on 2 September 2022, from https://data.worldbank.org.cn/indicator) is the source of information on technological advancement.

The variable description and descriptive statistics are shown in Table 1.

Table 1. Variable description and descriptive statistics.

Variable Types	Variable Abbreviation	Name	Definition	Unit
Explained variable	CO_2	Carbon dioxide	Carbon dioxide emissions from energy	Million tons equivalent
Explanatory variable	*EXPY*	Export sophistication of new energy industry	Weighted average sum of export sophistication of different new energy products in a country's new energy industry	USD
Control variables	*FDI*	Foreign direct investment	The Net FDI	100 Million USD
	Urb	Urbanization	The percentage of urban residents in the overall population	%
	IT	International Trade	The total amount of imports and exports	100 Million USD
Intermediary variable	*TP*	Technological progress	Cobb-Douglas production function	-
Variables	**Mean**	**St.Dev**	**Max**	**Min**
$LnCO_2$	5.523	1.345	9.261	3.290
LnEXPY	9.696	0.219	10.411	8.884
LnFDI	23.134	1.641	8.666	−1.405
LnIT	7.833	1.297	10.533	3.108
LnUrb	−0.347	0.274	0	−1.316
LnTP	6.570	0.871	7.717	4.024

Note: Descriptive statistical analysis is performed using the *tabstat* command in the stata15.0 software.

We gathered yearly panel data for 31 major economies worldwide from 1996 to 2021 based on data availability to confirm the impact of EXPY on CO_2. These 31 economies (as indicated in Table 2), which accounted for 48.67% of the global GDP in 2021, are made up of 22 developed countries or regions (subsequently referred to as the countries) and 9 emerging countries. More than 85% of the world's carbon emissions come from these 31 economies' total emissions of carbon dioxide, while their total new energy exports account for more than 95% of the world's new energy trading market. The chosen economies are adequate and representative. To reduce the potential heteroscedasticity of the sample data, we logarithmized all of the data.

Table 2. Summary of sample countries.

Sample Classification	Name of Economies
Developed countries	The United States (USA), Belgium (BEL), Germany (DEU), Canada (CAN), Austria (AUT), Switzerland (CHE), The Czech Republic (CZE), Denmark (DNK), Spain (ESP), Netherlands (NLD), France (FRA), Britain (GBR), Hong Kong (HKG), Hungary (HUN), Italy (ITA), Japan (JPN), Finland (FIN), Republic of Korea (KOR), Poland (POL), Portugal (PRT), Singapore (SGP), Sweden (SWE)
Developing countries	Brazil (BRA), Philippines (PHL), China (CHN), Thailand (THA), India (IND), Malaysia (MYS), Romania (ROU), Mexico (MEX), Russian Federation (RUS)

Before starting to apply the econometric model, it is necessary to test the stationarity of the original data and select the specific form of the model. This study will highlight the methodology in accordance with the following conceptual framework (see Figure 1 [51]).

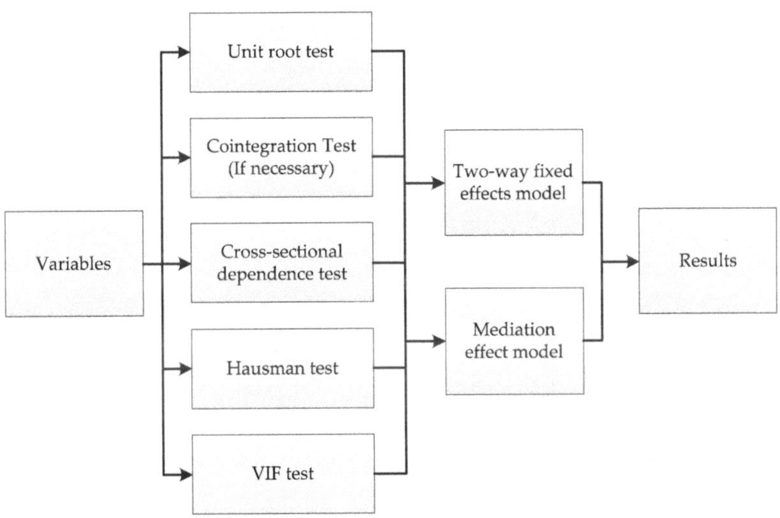

Figure 1. Conceptual framework.

4. Empirical Findings

When using the regression model to analyze the correlation between the explanatory variable and the explained variable, the phenomenon of pseudo-regression may occur, which means that the data of the explanatory variable and the explained variable is non-stationary, but the regression outcomes reveal that there is a statistical association between the two for some reason, and the regression results have no practical significance. To prevent pseudo-regression in the regression process, the original data must be tested for stationarity. The IPS test and Fisher test of the *xtunitroot* command are used to conduct a stationarity test on panel data; the results are presented in Table 3.

Table 3. The results of the unit root test.

	IPS Test		Fisher Test		Order of Integration
	Statistic	*p*-Value	Statistic	*p*-Value	
LnCO_2	−3.3720 ***	0.0004	5.1761 ***	0.0000	I(0)
Ln*EXPY*	−7.5256 ***	0.0000	14.1506 ***	0.0000	I(0)
Ln*FDI*	−10.0535 ***	0.0000	12.5743 ***	0.0000	I(0)
Ln*IT*	−1.6782 **	0.0467	9.2752 ***	0.0000	I(0)
Ln*Urb*	−5.1997 ***	0.0001	9.0213 ***	0.0000	I(0)
Ln*TP*	−5.1702 ***	0.0000	10.1308 ***	0.0000	I(0)

Note: ** $p < 0.05$, *** $p < 0.01$.

As shown in Table 3, the *p* values of the explained variable (LnCO_2), the explanatory variable (Ln*EXPY*), the control variables (Ln*Urb*, Ln*FDI* and Ln*IT*) and the intermediary variable (Ln*TP*) are all less than 0.05, rejecting the null hypothesis and accepting the alternative hypothesis, indicating that all variables are considered stationary.

In general, there are three varieties of panel models: fixed effects model, pool effect model, and random effect model. To ensure the validity and consistency of the estimated results of the regression model, it is necessary to identify the optimal model type based on the results of various tests. When comparing the fixed effect model with the pool effect model, the *xtcsd* command is used to assess the cross-section dependence of the panel data. The test statistic, 7.237, exceeds the critical value of 0.5811, which corresponds to a significance level of 1%. The initial assumption that there is no cross-section dependence is therefore refuted, and the model is regarded to have cross-section dependence. The *xtscc* command is then used to determine whether or not the model has individual effects. The test results indicate that the *p*-value is 0.000, allowing us to disapprove of the null hypothesis and assume that there are individual effects; therefore, the fixed effects model is superior to the pool effect model. The fixed effect model and the random effect model are commonly compared and chosen using the *Hausman* command. The test's findings show that the *p*-value is 0.000, failing to meet the 5% threshold for significance. Consequently, the initial hypothesis of the random effect model is refuted, showing that the fixed effects model is the preferable alternative. Combining the outcomes of the two comparisons, the two-way fixed effects model was subsequently applied to panel data regression.

Following model selection and the unit root test, the two-way fixed effects model (xtreg command for Stata 15.0) is used to examine the carbon emission effect of the new energy industry's export sophistication and the regression results are displayed in Table 4.

Table 4. Regression results of the baseline model.

	Model 1	Model 2	Model 3	Model 4
Ln*EXPY*	−0.320 *** (0.000)	−0.290 *** (0.000)	−0.210 *** (0.001)	−0.219 *** (0.000)
Ln*FDI*		0.046 *** (0.000)	0.028 *** (0.001)	0.029 *** (0.000)
Ln*IT*			0.244 *** (0.000)	0.117 *** (0.000)
Ln*Urb*				1.467 *** (0.000)
Country	Y	Y	Y	Y
Year	Y	Y	Y	Y
Constant	8.611 *** (0.000)	8.179 *** (0.000)	5.627 *** (0.000)	7.265 *** (0.000)
Mean VIF	-	1.01	2.20	1.96

Note: *p*-values in parentheses; *** $p < 0.01$.

This research uses the *vif* command to broaden the detection to guarantee that there is no multicollinearity across variables. The findings reveal that the VIF values of models 1 to 4 in Table 4 are both below 10, suggesting that there is no multicollinearity between variables.

According to the findings of the regression analysis, the correlation between Ln*EXPY* and LnCO_2 is less than 0, and the significance test is passed at the 1% level, indicating that enhancing the new energy industry's export sophistication will substantially reduce carbon dioxide emissions. Carbon dioxide emissions will drop by 0.219% for every percentage rise in Ln*EXPY*. The explanation for the negative inhibitory effect between LnEXPY and LnCO_2 is that as the new energy industry's export sophistication increases, the capital and technology content of the exported new energy commodities increases, and the demand for fossil energy for such capital- and technology-intensive commodities continues to decline. By optimizing the structure of energy consumption, carbon dioxide emissions are reduced.

At the 1% level of significance, the relationship between Ln*FDI* and LnCO_2 has an elasticity value of 0.029, which is statistically significant. Each 1% increase in net foreign investment will result in a 0.029% increase in carbon dioxide emissions. Although there may be a Pollution Halo effect of FDI on carbon emissions, empirical evidence suggests that FDI's Pollution Haven effect inevitably increases the host country's carbon emissions [52].

The elasticity coefficient between Ln*IT* and LnCO_2 emissions is 0.117, and it passed the 1% significance level test. The change of 1% in international trade will result in a change of 0.117% in carbon emissions. Promoting international trade, according to the principle of comparative advantage, would allow a country to develop goods with comparative advantages, lowering carbon emissions by boosting resource usage efficiency [53]. However, international trade-driven global economic growth has boosted demand for fossil fuels, resulting in rising global carbon emissions.

The positive impact of Ln*Urb* on LnCO_2 was tested at a significance level of 1%, indicating that urbanization has worsened carbon emissions despite the fact that urbanization could reduce carbon emissions through resource agglomeration and large-scale management [54,55]. However, increased urbanization also drives up the need for infrastructure and energy utilization, resulting in an increase in CO_2 [56]. The study's findings show that urbanization causes carbon emissions to grow at a faster rate than agglomeration causes them to decrease, with an increase in carbon dioxide emissions as a result.

Despite the fact that the panel regression results indicate that the new energy industry's export sophistication is conducive to reducing carbon emissions, it is necessary to employ a series of methods to ensure the conclusions' objectivity, and the results are given in Table 5.

Table 5. Summary of different robustness regression results.

	Model 5	Model 6	Model 7	Model 8
Ln*EXPY*	−0.140 *** (0.005)	−0.210 *** (0.003)	−0.141 ** (0.034)	−0.225 *** (0.000)
Ln*FDI*	0.026 *** (0.000)	0.028 *** (0.000)	0.033 *** (0.000)	0.027 *** (0.001)
Ln*TO*	0.123 *** (0.000)	0.156 *** (0.000)	0.133 *** (0.000)	0.126 *** (0.000)
Ln*Urb*	1.382 *** (0.000)	1.394 *** (0.000)	1.404 *** (0.000)	1.488 *** (0.000)
Ln*IS*				−0.253 *** (0.005)
Country	Y	Y	Y	Y
Year	Y	Y	Y	Y
Constant	2.819 *** (0.000)	6.869 *** (0.000)	6.331 *** (0.00)	6.965 *** (0.000)

Note: *p*-values in parentheses; ** $p < 0.05$, *** $p < 0.01$.

(1) Substitute the explained variable. Replace with the outlined variable. Model 5 shows the outcome of the robustness test using per capita carbon emissions rather than total emissions. The refitted regression result indicated a carbon reduction effect of the new energy industry's export sophistication, and the test was passed at the significance level of 1%. The regression coefficient symbols and significance for other variables are identical to the results of the standard regression. Overall, it can be concluded with confidence that improving *EXPY* can substantially reduce carbon emissions;

(2) Shrink the tail of explanatory variables. Due to the occurrence of singular values, there may be some variations between the regression estimate findings and the real scenario based on the derived explanatory factors. To avoid this situation, we use the fixed-effect model for panel regression and do a two-tailed treatment of 5% for the explanatory variables. The estimated coefficient of the ln$EXPY$ and lnCO_2 is −0.210 (see Model 6 in Table 5), suggesting that a 1% increase in ln$EXPY$ reduces carbon emissions by 0.210%. Other control variable regression coefficient symbols were consistent with the benchmark regression findings and passed the significance test, demonstrating the robustness of the benchmark regression results;

(3) Eliminate the interference of major international emergencies. Some unexpected large worldwide occurrences, such as the global subprime mortgage crisis in 2007 and the Corona Virus Disease 2019 (COVID-19), which caused varying degrees of recession in the export trade of major economies around the world, will have an effect on the estimates. In light of this, we delete data for a total of 5 years from 2007–2009 (the subprime mortgage crisis occurred in 2007 and ended in 2009) and 2020–2021 (COVID-19 occurred at the end of 2019 and rapidly evolved into a global event in early 2020) to eliminate the impact of these two major events on the regression results (as shown in Model 7). The correlation coefficient between ln$EXPY$ and lnCO_2 is less than zero, which is consistent with the benchmark regression findings. As a result, after controlling for big unexpected international events, the coefficient of the main independent variable is notably negative.

(4) Add a control variable. Taking into account the impact of missing variables, this paper controls the industrial structure variable and conducts panel regression once more. Model 8 shows that, after controlling for the industrial structure variable, the export sophistication of the new energy industry has a negative correlation with carbon dioxide emissions, and the other control variables' regression coefficients correspond to the benchmark regression. As a result, the carbon reduction effect of the new energy industry's export sophistication remains effective.

5. Further Discussion

5.1. Mechanism Inspection

The findings of the benchmark regression indicate a negative correlation between ln$EXPY$ and lnCO_2; however, additional research is required to determine how this relationship is mediated. According to some academics, rising export sophistication indicates that the export sector is advancing technologically, which indirectly raises a nation's overall technological level through active transmission or passive spillover. And technological progress can also significantly lower carbon emissions [57]. Exploring the potential role of technological progress as a mediator between export sophistication and carbon emissions is a crucial matter. Results of empirical regression using technological advancement as a study's mediator variable are shown in Figure 2.

As shown in Figure 2, the elasticity coefficients of Ln$EXPY$ and LnTP, as well as LnTP and lnCO_2, are all statistically significant at the 1% significance level, demonstrating that technological progress is one of the mechanisms by which the new energy industry's export sophistication affects carbon emissions. According to the change in coefficients, the direct effect of Ln$EXPY$ on lnCO_2 is −0.230, meaning that for every 1% increase in Ln$EXPY$, carbon dioxide emissions will decrease by 0.230%; however, with the intervention of technological progress, the total effect of Ln$EXPY$ on lnCO_2 is 0.219%. It's interesting to note that the new energy industry's export sophistication exhibits a negative relationship with technological progress in the mediated transmission process, i.e., an increase in the new energy industry's export sophistication will be detrimental to the domestic technological level. This is primarily because a country has a finite amount of innovation resources, and if it concentrates those resources on the new energy industry's export sophistication, it will exhaust those resources for domestic innovation.

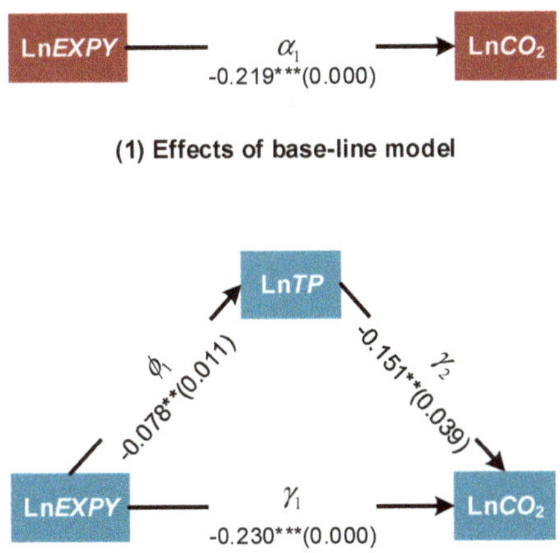

Figure 2. Analysis of moderator effects. Note: p-values in parentheses; ** $p < 0.05$, *** $p < 0.01$.

5.2. Heterogeneity Discussion

Given the vast differences in economic development between countries, the new energy industry's export sophistication may have various effects on carbon emission reduction in different countries. On this basis, we classified 31 sample countries according to their level of economic development into developed and less developed countries. The results of our investigation into the heterogeneity of the carbon emission reduction effects of the new energy industry's export sophistication at different economic development levels are presented in Figure 3.

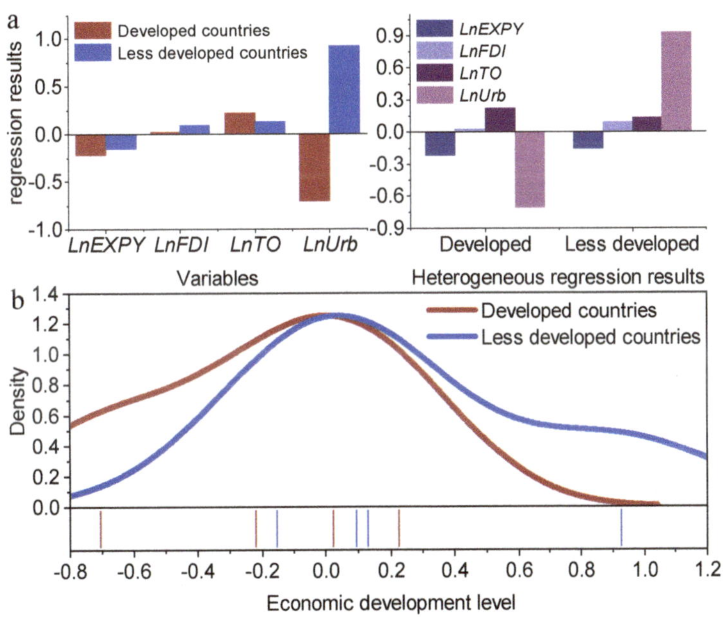

Figure 3. Heterogeneous regression outcomes for the effects of the new energy industry's export sophistication on carbon emissions. (**a**) represents the comparison of heterogeneity estimation coefficients, while (**b**) represents the distribution of lnEXPY coefficients estimated for different country types.

According to Figure 3, the elastic coefficient between Ln$EXPY$ and lnCO_2 in developed countries is −0.220, and it passes the 1% significance level test. While in less developed countries, CO_2 emissions will drop by 0.155% for every 1% improvement in Ln$EXPY$. The regression findings demonstrate that the new energy industry's export sophistication is helpful in lowering carbon emissions in both developed and developing countries. Furthermore, as compared to less developed countries, the new energy industry's export sophistication in developed countries has a greater influence on reducing carbon emissions. The explanation for this phenomenon is that economic development is the first priority for developing countries, and they prefer to continue consuming fossil energy rather than developing new energy for industries with high initial investment sunk costs, whereas developed countries prioritize environmental protection and are willing to invest heavily in the development of new energy industries and new energy technologies to achieve long-term goals.

6. Conclusions and Policy Implications

The optimization of energy consumption structure and the reduction of global carbon emissions are both greatly aided by the growth of the new energy sector. From the standpoint of export sophistication, this research investigates the direction, mechanism, and heterogeneity of the new energy industry's influence on carbon dioxide. To accomplish this, empirical experiments were conducted by gathering data from 1996 to 2021 from 31 of the world's major economies via the *UN Comtrade database, the World Bank Open Data*, and the *2022 BP Statistical Review of World Energy*. The findings indicate that the new energy industry's export sophistication may contribute to a decrease in carbon dioxide emissions, and this conclusion has withstood a number of robustness tests. The mechanism analysis reveals that the export sophistication of the new energy industry will have a crowding-out influence on domestic technological innovation, which is not conducive to achieving the global carbon emission reduction target. We also observe regional heterogeneity, as the effect of the new energy industry's export sophistication on carbon reduction is more pronounced in developed countries. In light of the significance of new energy in attaining carbon neutrality and a carbon peak, this research on the new energy industry provides a theoretical framework for the low-carbon transformation of the energy sector. This paper also provides evidence for the high-quality development of the new energy industry from the perspective of export sophistication, which is conducive to taking the initiative and the lead in the process of reshaping the global energy supply and demand pattern.

Based on the previous findings, this research proposes the three policy implications listed below.

Firstly, we should prioritize enhancing the new energy industry's export sophistication. Countries around the world should accumulate the production process of new energy products, actively enhance the production capacity of high-end new energy products, and cultivate their own international competitive advantage in the new energy industry. Secondly, innovation resources should be cultivated to mitigate the effect of export sophistication on domestic innovation resources being crowded out. In terms of the total amount of innovation resources, improve the training support for R&D personnel, and foster a group of scientific and technological innovators; In the development of the new energy industry, an additional new energy industry innovation fund will be established, which will be used for talent support and technological research and development in the new energy industry, and will increase support for the new energy industry. Finally, distinct new energy industry development plans should be developed, and the comparative advantages of various country types should be properly leveraged. Developed countries should speed up research into new energy utilization technologies, particularly those with zero carbon emissions, and accelerate the green energy transition. Developing countries should abandon the idea of development dependent on fossil fuels, lay out new energy products with comparative advantages, and gradually join the global new energy industry's international division of labor system.

It is important to note that this study is primarily based on the data from 31 of the world's major economies; however, if the countermeasures and suggestions in this study are used to guide the development of the new energy industry in a particular country, the effect may be greatly diminished due to the unique characteristics of the country. To overcome this limitation, future research will concentrate on a specific nation in order to devise countermeasures that are more compatible with the growth of the nation's new energy industry.

Author Contributions: Conceptualization, K.H. and T.W.; methodology, T.W.; software, T.W.; validation, T.W. and K.H.; formal analysis, T.W.; investigation, T.W.; resources, L.S.; data curation, T.W.; writing—original draft preparation, T.W., K.H. and J.P.; writing—review and editing, T.W., K.H., J.P. and L.S.; visualization, J.P.; supervision, L.S.; project administration, K.H.; funding acquisition, K.H. All authors have read and agreed to the published version of the manuscript.

Funding: This research was funded by the National Natural Science Foundation of China, grant number 71904178; the MOE (Ministry of Education in China) Post-supported Projects of Philosophy and Social Sciences Research, grant number 22JHQ063; and the MOE (Ministry of Education in China) Project of Humanities and Social Sciences research project of Science, grant number 19YJA630062 and 22YJC630107.

Data Availability Statement: Not applicable.

Conflicts of Interest: The authors declare no conflict of interest.

Abbreviations

CO_2	Carbon dioxide
PGDP	Per capita GDP
EXPY	The export sophistication of the new energy industry
FDI	Foreign direct investment
IT	International trade
Urb	Urbanization
TP	Technological progress

References

1. Sinha, A. Inequality of renewable energy generation across OECD countries: A note. *Renew. Sustain. Energy Rev.* **2017**, *79*, 9–14. [CrossRef]
2. Moriarty, P.; Honnery, D. Can renewable energy power the future? *Energy Policy* **2016**, *93*, 3–7. [CrossRef]
3. Ferdinand, P. Westward ho-the China dream and 'one belt, one road': Chinese foreign policy under Xi Jinping. *Int. Aff.* **2016**, *92*, 941–957. [CrossRef]
4. Central People's Government of the People's Republic of China. Available online: http://www.gov.cn/xinwen/2023-02/14/content_5741481.htm (accessed on 25 February 2023).
5. Wang, Z.X.; Zheng, H.H.; Pei, L.L.; Jin, T. Decomposition of the factors influencing export fluctuation in China's new energy industry based on a constant market share model. *Energy Policy* **2017**, *109*, 22–35. [CrossRef]
6. Wang, Z.J.; Jebli, M.B.; Madaleno, M.; Doğan, B.; Shahzad, U. Does export product quality and renewable energy induce carbon dioxide emissions: Evidence from leading complex and renewable energy economies. *Renew. Energy* **2021**, *171*, 360–370. [CrossRef]
7. Bhattacharya, M.; Churchill, S.A.; Paramati, S.R. The dynamic impact of renewable energy and institutions on economic output and CO_2 emissions across regions. *Renew. Energy* **2017**, *111*, 157–167. [CrossRef]
8. Edziah, B.K.; Sun, H.P.; Adom, P.K.; Wang, F.; Agyemang, A.O. The role of exogenous technological factors and renewable energy in carbon dioxide emission reduction in Sub-Saharan Africa. *Renew. Energy* **2022**, *196*, 1418–1428. [CrossRef]
9. Wang, Q.; Guo, J.L.; Li, R.R.; Jiang, X.T. Exploring the role of nuclear energy in the energy transition: A comparative perspective of the effects of coal, oil, natural gas, renewable energy, and nuclear power on economic growth and carbon emissions. *Environ. Res.* **2023**, *221*, 115290. [CrossRef]
10. Dong, K.Y.; Hochman, G.; Zhang, Y.Q.; Sun, R.J.; Li, H.; Liao, H. CO_2 emissions, economic and population growth, and renewable energy: Empirical evidence across regions. *Energy Econ.* **2018**, *75*, 180–192. [CrossRef]
11. Acheampong, A.O.; Adams, S.; Boateng, E. Do globalization and renewable energy contribute to carbon emissions mitigation in Sub-Saharan Africa? *Sci. Total Environ.* **2019**, *677*, 436–446. [CrossRef]

12. Habiba, U.; Cao, X.B.; Anwar, A. Do green technology innovations, financial development, and renewable energy use help to curb carbon emissions? *Renew. Energy* **2022**, *193*, 1082–1093. [CrossRef]
13. Rahman, M.M.; Alam, K. Effects of corruption, technological innovation, globalisation, and renewable energy on carbon emissions in asian countries. *Util. Policy* **2022**, *79*, 101448. [CrossRef]
14. Djellouli, N.; Abdelli, L.; Elheddad, M.; Ahmed, R.; Mahmood, H. The effects of non-renewable energy, renewable energy, economic growth, and foreign direct investment on the sustainability of African countries. *Renew. Energy* **2022**, *183*, 676–686. [CrossRef]
15. Al-Mulali, U.; Saboori, B.; Ozturk, I. Investigating the environmental Kuznets curve hypothesis in Vietnam. *Energy Policy* **2015**, *76*, 123–131. [CrossRef]
16. Zaidi, S.A.H.; Hou, F.; Mirza, F.M. The role of renewable and non-renewable energy consumption in CO_2 emissions: A disaggregate analysis of Pakistan. *Environ. Sci. Pollut. Res.* **2018**, *25*, 31616–31629. [CrossRef]
17. Jebli, M.B.; Youssef, S.B. The role of renewable energy and agriculture in reducing CO_2 emissions: Evidence for North Africa countries. *Ecol. Indic.* **2017**, *74*, 295–301. [CrossRef]
18. Michaely, M. *Trade, Income Levels, and Dependence*; Elsevier: Amsterdam, The Netherlands, 1984.
19. Hausmann, R.; Hwang, J.; Rodrik, D. What you export matters. *J. Econ. Growth* **2007**, *12*, 1–25. [CrossRef]
20. Jarreau, J.; Poncet, S. Export sophistication and economic growth: Evidence from China. *J. Dev. Econ.* **2012**, *97*, 281–292. [CrossRef]
21. Rehman, F.U.; Islam, M.M.; Raza, S.A. Does disaggregate energy consumption matter to export sophistication and diversification in OECD countries? A robust panel model analysis. *Renew. Energy* **2023**, *206*, 274–284. [CrossRef]
22. Su, X.Y.; Anwar, S.; Zhou, Y.; Tang, X. Services trade restrictiveness and manufacturing export sophistication. *N. Am. J. Econ. Financ.* **2020**, *51*, 101058. [CrossRef]
23. Song, Y.G.; Hao, X.Z.; Zheng, L. Intermediate import, independent innovation and export sophistication of Chinese manufacturing enterprises. *Struct. Chang. Econ. D* **2022**, *60*, 126–140. [CrossRef]
24. Zheng, H.H.; Wang, Z.X. Measurement and comparison of export sophistication of the new energy industry in 30 countries during 2000–2015. *Renew. Sust. Energy Rev.* **2019**, *108*, 140–158. [CrossRef]
25. Cao, X.P.; Zhu, F.X.; Wang, R. Research on the Dynamic Evolution of Technical Structure of New Energy Industry Export. *Forum Sci. Technol. China* **2019**, *7*, 76–83. (In Chinese)
26. Liu, Z.; Guan, D.; Wei, W.; Davis, J.S.; Ciais, P.; Bai, J.; Peng, S.S.; Zhang, Q.; Hubacek, K.; Marland, G.; et al. Reduced carbon emission estimates from fossil fuel combustion and cement production in China. *Nature* **2015**, *524*, 335–338. [CrossRef]
27. Worrell, E.; Price, L.; Martin, N.; Hendriks, C.; Meida, O.L. Carbon dioxide emissions from the global cement industry. *Annu. Rev. Energy Env.* **2001**, *26*, 303–329. [CrossRef]
28. Sun, Z.R.; Liu, Y.D.; Yu, Y.N. China's carbon emission peak pre-2030: Exploring multi-scenario optimal low-carbon behaviors for China's regions. *J. Clean. Prod.* **2019**, *231*, 963–979. [CrossRef]
29. Ridzuan, N.A.M.; Marwan, N.F.; Khalid, N.; Ali, H.M.; Tseng, M.L. Effects of agriculture, renewable energy, and economic growth on carbon dioxide emissions: Evidence of the environmental Kuznets curve. *Resour. Conserv. Recycl.* **2020**, *160*, 104879. [CrossRef]
30. Sufyanullah, K.; Ahmad, K.A.; Ali, M.A.S. Does emission of carbon dioxide is impacted by urbanization? An empirical study of urbanization, energy consumption, economic growth and carbon emissions-Using ARDL bound testing approach. *Energy Policy* **2022**, *164*, 112908. [CrossRef]
31. Raihan, A. The dynamic nexus between economic growth, renewable energy use, urbanization, industrialization, tourism, agricultural productivity, forest area, and carbon dioxide emissions in the Philippines. *Energy Nexus* **2023**, *9*, 100180. [CrossRef]
32. Lu, W.C.; Kasimov, I.; Saydaliev, H.B. Foreign direct investment and renewable energy: Examining the environmental Kuznets curve in resource-rich transition economies. *Renew. Energy* **2023**, *208*, 301–310.
33. Wang, T.; Peng, J.C.; Wu, L. Heterogeneous effects of environmental regulation on air pollution: Evidence from China's prefecture-level cities. *Environ. Sci. Pollut. Res.* **2021**, *28*, 25782–25797. [CrossRef]
34. Baron, R.M.; Kenny, D.A. The moderator–mediator variable distinction in social psychological research: Conceptual, strategic, and statistical considerations. *J. Pers. Soc. Psychol.* **1986**, *51*, 1173. [CrossRef] [PubMed]
35. Lu, H.F.; Peng, J.C.; Lu, X.Y. Do Factor Market Distortions and Carbon Dioxide Emissions Distort Energy Industry Chain Technical Efficiency? A Heterogeneous Stochastic Frontier Analysis. *Energies* **2022**, *15*, 6154. [CrossRef]
36. Qian, L.; Xu, X.L.; Sun, Y.; Zhou, Y.J. Carbon emission reduction effects of eco-industrial park policy in China. *Energy* **2022**, *261*, 125315. [CrossRef]
37. Shi, B.B.; Li, N.; Gao, Q.; Li, G.Q. Market incentives, carbon quota allocation and carbon emission reduction: Evidence from China's carbon trading pilot policy. *J. Environ. Manag.* **2022**, *319*, 115650. [CrossRef]
38. Fan, P.S. Export Technological Sophistication of China: Measurement and Impact Factor. *Discrete Dyn. Nat. Soc.* **2021**, *2021*, 1–9. [CrossRef]
39. Cole, M.A. Trade, the pollution haven hypothesis and the environmental Kuznets curve: Examining the linkages. *Ecol. Econ.* **2004**, *48*, 71–81. [CrossRef]
40. Zhou, Y.; Fu, J.T.; Kong, Y.; Wu, R. How foreign direct investment influences carbon emissions, based on the empirical analysis of Chinese urban data. *Sustainability* **2018**, *10*, 2163. [CrossRef]
41. Birdsall, N.; Wheeler, D. Trade policy and industrial pollution in Latin America: Where are the pollution havens? *J. Environ. Dev.* **1993**, *2*, 137–149. [CrossRef]

42. Mert, M.; Caglar, A.E. Testing pollution haven and pollution halo hypotheses for Turkey: A new perspective. *Environ. Sci. Pollut. Res.* **2020**, *27*, 32933–32943. [CrossRef]
43. Ali, U.; Guo, Q.; Nurgazina, Z.; Sharif, A.; Kartal, T.M.; Depren, K.S.; Khan, A. Heterogeneous impact of industrialization, foreign direct investments, and technological innovation on carbon emissions intensity: Evidence from Kingdom of Saudi Arabia. *Appl. Energy* **2023**, *336*, 120804. [CrossRef]
44. Shahbaz, M.; Nasreen, S.; Ahmed, K.; Hammoudeh, S. Trade openness–carbon emissions nexus: The importance of turning points of trade openness for country panels. *Energy Econ.* **2017**, *61*, 221–232. [CrossRef]
45. Chen, Y.L.; Wang, Z.; Zhong, Z.Q. CO_2 emissions, economic growth, renewable and non-renewable energy production and foreign trade in China. *Renew. Energy* **2019**, *131*, 208–216. [CrossRef]
46. Anochiwa, L.I.; Agbanike, T.F.; Chukwu, A.B.; Ikpe, M.; Otta, N.N. Urbanization and carbon emissions: Looking at the role of mobile phone adoption in Sub-Saharan African countries. *Environ. Sci. Pollut. Res.* **2022**, *29*, 78526–78541. [CrossRef]
47. Guo, X.M.; Fang, C.L. How does urbanization affect energy carbon emissions under the background of carbon neutrality? *J. Environ. Manage* **2023**, *327*, 116878.
48. Wang, W.Z.; Liu, L.C.; Liao, H.; Wei, Y.M. Impacts of urbanization on carbon emissions: An empirical analysis from OECD countries. *Energy Policy* **2021**, *151*, 112171. [CrossRef]
49. Cai, J.Y.; Zheng, H.Y.; Vardanyan, M.; Shen, Z.Y. Achieving carbon neutrality through green technological progress: Evidence from China. *Energy Policy* **2023**, *173*, 113397. [CrossRef]
50. Cobb, C.W.; Douglas, P.H. A theory of production. *Am. Econ. Rev.* **1928**, *18*, 139–165.
51. Fuinhas, J.A.; Koengkan, M.; Leitão, N.C.; Nwani, C.; Uzuner, G.; Dehdar, F.; Relva, S.; Peyerl, D. Effect of Battery Electric V ehicles on Greenhouse Gas Emissions in 29 European Union Countries. *Sustainability* **2021**, *13*, 13611. [CrossRef]
52. Zhang, Y.; Zhang, S.F. The impacts of GDP, trade structure, exchange rate and FDI inflows on China's carbon emissions. *Energy Policy* **2018**, *120*, 347–353. [CrossRef]
53. Shahbaz, M.; Gozgor, G.; Hammoudeh, S. Human capital and export diversification as new determinants of energy demand in the United States. *Energy Econ.* **2019**, *78*, 335–349. [CrossRef]
54. Adams, S.; Nsiah, C. Reducing carbon dioxide emissions; Does renewable energy matter? *Sci. Total Environ.* **2019**, *693*, 133288. [CrossRef] [PubMed]
55. Dou, Y.; Zhao, J.; Malik, M.N.; Dong, K.Y. Assessing the impact of trade openness on CO_2 emissions: Evidence from China-Japan-ROK FTA countries. *J. Environ. Manag.* **2021**, *296*, 113241. [CrossRef]
56. Kasman, A.; Duman, Y.S. CO_2 emissions, economic growth, energy consumption, trade and urbanization in new EU member and candidate countries: A panel data analysis. *Econ. Model* **2015**, *44*, 97–103. [CrossRef]
57. Leitão, J.; Ferreira, J.; Santibanez-González, E. New insights into decoupling economic growth, technological progress and carbon dioxide emissions: Evidence from 40 countries. *Technol. Forecast Soc. Chang.* **2022**, *174*, 121250. [CrossRef]

energies

Article

Assessing the Implications of Ecological Civilization Pilots in Urban Green Energy Industry on Carbon Emission Mitigation: Evidence from China

Peng Zhang [1], Lei Tan [2] and Fei Liu [3,*]

[1] School of Economics and Management, Xi'an Aeronautical Institute, Xi'an 710077, China; 201204001@xaau.edu.cn
[2] School of Public Administration, Northwest University, Xi'an 710127, China; tanlei@stumail.nwu.edu.cn
[3] School of Philosophy, Northwest University, Xi'an 710127, China
* Correspondence: liufei@nwu.edu.cn

Abstract: This study aims to explore the role of China's Ecological Civilization Pilot Policies in carbon emissions reduction within the urban green energy industry. It further investigates how these policies influence carbon emissions. To achieve this, a unique incentive–constraint model is established considering China's distinctive political system. The DID model was used in this study, employing Chinese city data spanning from 2009 to 2020 and analyzing urban panel data with the use of two specific policies as quasi-natural experiments. The analysis reveals the following key findings: (i) Ecological Civilization Pilot Policies in the energy industry substantially contribute to carbon emission reduction through the effects of technological progress and industrial structure optimization; (ii) the unique incentive–restraint mechanism within these policies enhances their effectiveness, with short-term incentives and carefully designed assessment criteria playing a pivotal role in their successful implementation. These findings carry substantial implications for shaping environmental policies within the energy industry, emphasizing the importance of such policies in the ongoing global effort to reduce carbon emissions and promote sustainability.

Keywords: carbon emissions; ecological civilization; incentive–constraint mechanism

Citation: Zhang, P.; Tan, L.; Liu, F. Assessing the Implications of Ecological Civilization Pilots in Urban Green Energy Industry on Carbon Emission Mitigation: Evidence from China. *Energies* **2023**, *16*, 7638. https://doi.org/10.3390/en16227638

Academic Editors: Jiachao Peng, Le Wen, Jianzhong Xiao, Ming Yi and Mingyue (Selena) Sheng

Received: 27 October 2023
Revised: 14 November 2023
Accepted: 15 November 2023
Published: 17 November 2023

1. Introduction

Pilot policies represent a distinctive policy, particularly when applied to the experimental implementation of "ecological civilization", a rarity on the global stage. The notion of ecological civilization constitutes a unique buzzword within China, having garnered substantial influence within the nation since its integration into the ideology of the Communist Party of China in 2007 [1]. While this concept's definition bears a resemblance to that of "Ecological democracy" [2], China's developmental trajectory has transformed it into a societal vision that places paramount importance on enhancing the well-being of the populace and advancing national development with a steadfast commitment to the principles of sustainable development [3]. Since 2013, the Chinese government has implemented Ecological Civilization Pilot (ECP) policies in energy industry fields which encompass various strategies such as upgrading the energy structure, developing green industries, and protecting the ecological environment. These policies have provided China with practical foundations and accumulated experience in reducing carbon emissions and taking climate action [4].

The primary objective of this study is to delve into the multifaceted impact of China's ECP policies, implemented over a decade, on the reduction of regional carbon emissions. The interplay between these policies and regional carbon emission patterns is scrutinized, taking into consideration the unique incentive–restraint mechanisms that influence local officials. Empirical insights into the effectiveness of the ECP policy paradigm and how incentive and constraint mechanisms impact policy outcomes are sought after in this research.

Our research extends beyond the validation of existing theories, seeking to provide practical insights that have broader implications for global governance. The ongoing debate between neoclassical economics and the Porter Hypothesis, particularly within the context of environmental regulatory policies and their impact on carbon emissions, is navigated [5–7]. The distinctions in policy intensity, objectives, mechanisms, and assessments among various environmental regulatory policies are focused on, shedding light on the nuances of government policies and their varying effects. Furthermore, the potential to inform future policy formulation and promotion exists in our findings. The relationship between ECP policies, incentive–constraint mechanisms, and policy effects is examined, contributing to a theoretical foundation that can guide policymakers and researchers in their endeavors to address carbon emissions and ecological civilization.

The remainder of this paper proceeds as follows: Section 2 provides an overview of the theoretical underpinnings and establishes a hypothesis that will be utilized in the empirical examination. Section 3 describes the data and methodology. Section 4 provides the core results on the impact of ECP policies and discusses how incentive–constraint mechanisms influence these policies. Section 5 concludes this paper.

2. Theoretical Framework

2.1. ECP Policies Overview

In December 2013, China's National Development and Reform Commission (NDRC) proposed the establishment of an "Ecological Civilization Demonstration Area" in 100 regions across China. In 2017, China initiated the development of "China's Demonstration Cities & Counties for Ecological Civilization Construction". Both pilot policies were implemented at the prefecture-level cities and exhibited significant differences in policy indicators. This divergence allows for an impartial analysis of the correlation between ECP policies and carbon emissions. For the purpose of experimentation, these policies will be referred to as Policy A1 and Policy A2, respectively. The specific details of these two policies can be found in Table 1.

Table 1. ECP Policies.

Policy Details	Policy A1	Policy A2
Policy Name	China's Demonstration Cities & Counties for Ecological Civilization Construction.	Ecological Civilization Demonstration Area
Policy Objective	Establish a model for ecological civilization construction.	Explore ecological civilization construction in pilot areas.
Entry Method	Application Review Process.	Application Review Process
Responsible Agency	Ministry of Environmental Protection.	National Development and Reform Commission, Ministry of Finance, Ministry of Land and Resources, Ministry of Water Resources, Ministry of Agriculture, State Forestry Administration.
Policy Characteristics	National-level green honor.	Selective experimentation.
Policy Indicators	Clear construction targets and management procedures.	Pilot areas determine policy indicators based on their local conditions.
Assessment Method	Assessment is conducted first and the title is awarded after passing the assessment.	Regular inspections with qualification cancellation for areas that fail acceptance after the five-year construction period.
Policy Coverage	Titles are awarded annually, covering a total of 262 cities and counties in 2020.	In 2014 and 2015, a total of 100 representative areas were selected.

Analyzing the content of the pilot policies reveals commonalities between Policy A1 and Policy A2. Both policies feature an application–review mechanism, whereby local governments are required to initiate the application process, culminating in central authorities determining the approved list following a comprehensive review. Furthermore, both policies entail the establishment of specific construction and assessment criteria.

Notably, these policies diverge in terms of their respective incentive assessment mechanisms, manifesting as follows:

1. Incentive effects: Policy A1 exhibits a stronger actual incentive impact. As a form of regional "green honor", Policy A1 represents a short-term honor. Local governments need to meet the nationally prescribed assessment criteria within a short time frame. Afterward, the Ministry of Ecology and Environment issues corresponding titles to regions that pass the assessment. These titles can be considered a local official's green achievements. Therefore, local leaders strive to meet the assessment standards. In contrast, Policy A2, which involves long-term pilot programs, is influenced by the actions of local leaders. Currently, implementing this policy does not necessarily guarantee incentives for local officials. Furthermore, the central government has not specified how honors can be obtained under Policy A2, potentially leading local officials to lack motivation for such pilot programs.
2. Constraint systems: Policy A1, serving as an "honorary title" for a city, has stricter evaluations due to clear assessment criteria. This process involves a one-time assessment; once the criteria are met, the title is conferred without subsequent evaluation phases. As a result, this policy may cause local governments to prioritize their economic interests and discontinue the specific ecological civilization development measures outlined during the application and approval process. On the other hand, Policy A2 is an exploratory pilot, where the assessment criteria are unclear. Yet, there is a follow-up evaluation mechanism, and local governments must face the possibility of forced withdrawal if they do not pass the assessment.

When considering the two policies comprehensively, it becomes apparent that Policy A1 exhibits higher policy intensity. This is because local governments are obligated to fulfill tasks assigned by the central government in accordance with established regulations before undergoing central government inspections. This aligns with the principles of the administrative subcontracting system and the theories related to promotion tournaments. In contrast, Policy A2 involves a broader range of departments in initiating pilot programs and allows for more discretion. Local governments under Policy A2 may not have a greater sense of urgency to complete these tasks compared to those under Policy A1. There are some areas where both policies are concurrently implemented. In these cities, local governments must ensure the effective implementation of Policy A1 to earn the honorary title for their cities while also meeting the objectives of Policy A2 to address certain issues. Consequently, it can be inferred that among all cities, those implementing both policies exhibit higher policy intensity compared to those implementing either policy individually.

2.2. *Impact of ECP Policies on Carbon Emissions*

We believe that the implementation of ECP policies has led to higher requirements for energy enterprises. To align with the objectives delineated by the central government, local administrations operating within the framework of the target responsibility system will necessitate a transformation in their governance approach. This transition entails the adoption of more stringent environmental regulations within their respective jurisdictions. Typically, these measures fall into two primary categories: incentives and punishments [8,9]. Incentives encourage enterprises to reduce carbon emissions through diverse methods, encompassing technological advancements and production reduction, thereby catalyzing a city-wide reduction in carbon emissions. Conversely, punishments intensify the consequences of polluting emissions, discouraging polluting companies from continuing environmentally harmful production practices solely for economic gain. This ultimately leads to a reduction in carbon emissions. In this scenario, the implementation of ECP policies requires energy enterprises to augment their investments in research and development (R&D) and innovation. This augmentation aims to cultivate environmentally sustainable and more efficient technologies. As a result, it reduces energy and resource waste, facilitating proficient production management and cost-saving measures. Consequently, this initiative culminates in the reduction of carbon emission intensity, an enhancement of carbon efficiency, and the promotion of sustainable development.

Hypothesis 1 (H1). *The implementation of ECP policies incentivizes urban areas to reduce carbon intensity and enhance carbon efficiency in the energy industry.*

2.2.1. Energy Industrial Optimization

ECP policies typically align with the long-term carbon reduction objectives of the nation or region. When formulating these policies, the central government establishes the level of carbon emissions within the region. Local governments, upon receiving these tasks, continuously employ administrative measures to compel enterprises to align their actions with the long-term policy objectives. These environmental regulatory measures compel both low-end and high-energy-consuming enterprises to undertake industrial upgrades, thereby reducing pollution emissions during the production process [10,11]. Consequently, they optimize the industrial structure, expedite regional industrial transformation and enhancement, reduce carbon intensity, and enhance carbon efficiency.

The requirements for ecological environment development will lead the government to increase the costs of environmental governance within the government budget while implementing environmental regulatory policies. This will also result in corresponding expenditures on green public services [12,13], thereby promoting the improvement of the ecological environment in the region. The "pollution haven" theory suggests that differences in environmental regulations among various regions can impact the decision-making process of pollution-intensive industries [14]. Regions with strict environmental regulations may incur higher costs for energy enterprises due to environmental issues, which can result in increased production costs and reduced comparative advantages for their products. Conversely, regions with more relaxed environmental regulations may attract polluting enterprises by offering lower environmental costs, thus becoming "pollution havens". Enterprises in the energy sector are categorized as being subject to stringent environmental regulations, consequently incurring significant costs associated with ecological and environmental management, thereby resulting in elevated production expenses. As the implementation of ecological civilization initiatives gains traction in a specific region, the corresponding policy imperatives stimulate improvements in the local ecological environment. This, in turn, creates a favorable external environment for the development of green technology-oriented industries, effectively mitigating the costs associated with industrial transformation. Concurrently, these policy-driven changes compel polluting enterprises to either exit the region or undergo significant industrial upgrades. This dynamic fosters a competitive environment in which regions strive to surpass each other in terms of environmental performance, resembling a "race to the top". Ultimately, this leads to a significant increase in the regional industrial agglomeration effect, thereby reducing carbon intensity and improving carbon efficiency.

Hypothesis 2a (H2a). *The implementation of ECP policies incentivizes urban areas to reduce carbon intensity and enhance carbon efficiency by optimizing energy industrial structure.*

2.2.2. Green Technological Progress

In the domain of technological progress, the implementation of ECP policies has a significant impact on the extent of environmentally friendly innovation in the area, primarily driven by two contrasting effects: the "compliance cost" and the "innovation compensation".

The "compliance cost" signifies how ECP policies prompt local governments to recalibrate their expectations concerning the ecological environment. As these policies are enacted, local governments enhance relevant ecological standards and emphasize corporate behavior regarding energy conservation and emission reduction. This increases the "compliance cost" for businesses, resulting in a "crowding out effect" that displaces innovation inputs, leading to the outflow of capital and hindering technological advancement [15,16]. While the "compliance cost" compels enterprises to either transition or upgrade their industries, it simultaneously hinders the improvement of carbon efficiency within the region.

The concept of "innovation compensation" refers to the practice in which local governments provide financial incentives to businesses that achieve green innovation in line with the predefined objectives of the central government. Following the "Porter Hypothesis", suitable environmental regulations induce businesses to internalize costs, thus propelling them to actively devise green processes, products, and technologies. By doing so, companies can not only alleviate the financial burden of policy implementation but also potentially generate additional revenue [17]. Initiated from the perspective of public choice theory at the enterprise level, this "innovation compensation" for green technological progress can rouse the subjective initiative of enterprises functioning as "rational economic agents". This stimulation encourages them to invest in green technology innovation, leading to improved efficiency in the use of raw materials and energy, reduced operational costs, and the attainment of policy incentives [18]. Ultimately, this contributes to the achievement of environmental regulatory objectives, such as the reduction of carbon emissions and the enhancement of carbon efficiency.

Hypothesis 2b (H2b). *The implementation of ECP policies incentivizes urban areas to reduce carbon intensity and enhance carbon efficiency by promoting technological progress.*

2.3. Incentive–Restraint Mechanism in ECP Policy

While countries worldwide have established environmental laws and policies [19–21], China's ECP policy stands out for its distinctive political context. China's current implementation of ECP policies predominantly adheres to a "local application, central oversight" model. Drawing from the theoretical perspective of "administrative subcontract" as expounded by Zhou [22], China's governance model can be synthesized as a composite of "vertical subcontracting" and "horizontal competition". The central government gradually delegates public administrative responsibilities and administrative discretion to subnational levels, empowering local administrative leaders with substantial authority. Concurrently, the power to appoint lower-level government leaders is vested in their higher-level "contractors". This model encapsulates the gradual devolution of public administrative responsibilities and administrative discretion to local tiers, endowing local administrative leaders with significant influence. However, in the process of policy issuance at different levels, local governments, acting as both "agents" responsible for policy implementation and "self-interested" entities, may not always align their goals with those of the central government. This duality stems from local governments serving as agents for superior governments' directives while concurrently pursuing their own political and financial interests.

To address this principal–agent problem between central and local governments, the central government must establish incentive and restraint mechanisms. We believe that this unique Chinese incentive–restraint mechanism impacts two opposing effects: "compliance cost" and "innovation compensation". Evaluating how incentives and constraints affect carbon emissions and carbon efficiency under different policies is imperative for comprehending China's pilot policies.

2.3.1. Incentive Mechanism

Local government officials' career trajectories are often intricately linked to opportunities for advancement. In the Chinese political system, there is a strong emphasis on performance-oriented promotions, with outstanding achievements being a prerequisite for ascending to higher-level government positions. As a result, local officials are often required to demonstrate their worth through notable accomplishments in order to stand out in the competition for government positions. In recent years, China has increasingly prioritized the development of a "green economy" and the pursuit of "dual carbon goals". For local leaders, reducing carbon emissions has now become one of the key achievements they must strive for. In regions where ECP policies are implemented, local governments acting as "self-interested actors" actively engage in ecological civilization construction. They diligently adhere to established objectives and compete with other local governments to gain recognition from the central government, achieve political victories, and obtain political incentives [23]. At this point, local governments invest more effort in ECP projects, mobilize resources to enhance the "innovation compensation" effect, and stimulate regional enterprises to have a greater innovation drive.

However, incentives do not necessarily lead to good results [24]. Empirical evidence indicates that high-ranking local officials in China typically hold their positions for a duration of 3–5 years. In order to achieve quick results, these officials tend to prioritize policies with short-term objectives. In contrast, leadership transitions may occur during the extended policy period in contexts defined by long-term objectives. The political legacy of the departing official is inherited by their successor, which can reduce the motivation of local officials. As a result, local authorities may prefer a strategy that emphasizes short-term incentives.

In this context, we believe that local officials are more inclined to drive policies aimed at reducing carbon emissions when short-term incentives are at play. Conversely, in regions with longer-term incentives, local officials may lack the motivation to drive policy implementation. This results in policies that help enterprises generate an "innovation compensation" but fail to offset "compliance costs". Consequently, "compliance costs" exceed "innovation compensation", leading to a reduction in carbon intensity but not an improvement in carbon efficiency.

Hypothesis 3a (H3a). *Compared to long-term incentives, ECP policies with short-term incentives are more effective in reducing carbon intensity and improving carbon efficiency through reasonable incentive mechanisms.*

2.3.2. Constraint Mechanism

Throughout the policy implementation process, local governments serving as "agents" entrusted with specific responsibilities often find themselves having to complete multiple tasks assigned by the central government simultaneously. Consequently, when conflicts of interest emerge between the central and local governments, local governments may utilize their discretion, granted by the central government, to adopt different strategies for implementing pilot policies based on the level of alignment of interests and the pressure to execute. This is often referred to as a "differential coping" strategy. In the absence of effective supervision, even though local governments, in their role as "agents", may actively engage in the application of pilot policies in response to the central government they may lack the sustained motivation to persistently and actively carry out these policy experiments [25]. Consequently, they will not enact stringent environmental regulatory measures, thus failing to impact carbon emissions within their respective regions. To regulate the "differential coping" strategy of local governments, the central government needs to establish certain constraint mechanisms, whether they be in the form of assessment systems or penalty systems, to standardize the behavior of local governments and prevent any potential negligence.

So, it is crucial that the stipulations of these constraint mechanisms are rational. Otherwise, in regions subjected to higher levels of constraint, while the central government's assessments might deter local governments from inaction to some extent, excessive levels of constraint may induce instances of data concealment by local authorities. Several scholars have explored the phenomenon of firms opting for quantity innovation over quality innovation [26]. The results of these studies indicate that when faced with innovation incentives companies tend to actively increase the number of patent applications. However, this growth primarily focuses on the "quantity" rather than the "quality" of innovation. As a result, there is an increase in non-invention patents, but no significant improvement in technology or product quality is achieved.

When environmental regulatory policies become excessively stringent, local officials may resort to strategic innovation. This response arises from the adverse impact of overly rigorous environmental regulations on the production and operations of local businesses, resulting in heightened operational costs. In their efforts to alleviate these costs and maintain the performance of local governments, local officials may choose innovations that demonstrate superficial compliance with regulations while lacking substantive environmental improvements. Such innovations may encompass compliance-focused pollution control measures, which may not necessarily lead to significant environmental enhancements. This means that local officials may prioritize performance and business operations over genuine efforts to drive sustainable environmental improvements.

In the broader context, it is essential to strike a balance between innovation incentives and regulatory constraints. An overemphasis on either side can result in adverse consequences. Within the framework of China's political landscape, when constraint intensity is reasonably calibrated, ECP policies can encourage local officials to focus on substantial innovation. This entails implementing significant measures to drive technological innovation and motivating enterprises to attain "innovation compensation" surpassing their "compliance costs". However, if constraint intensity is excessively high, local officials, acting as intermediaries, may opt for strategic innovation due to the disproportionate effort-to-reward ratio. This choice is made to maintain their political performance and ensure the stable operation of enterprises. In such scenarios, even though local governments, acting as "agents" for the central government, might still actively participate in the application of relevant pilot policies, due to the unreasonable constraint setting they lack the motivation to continue active policy experimentation. This, in turn, hinders the implementation of robust environmental regulatory measures, resulting in the ineffectiveness of both "compliance costs" and "innovation rewards" throughout this process.

Hypothesis 3b (H3b). *As the constraint intensity of ECP policies transitions from weak to strong, local governments tend to favor strategic innovation over substantive innovation.*

3. Materials and Methods

3.1. DID Model with Multiple Periods

The ECP policies can be regarded as a quasi-natural experiment in which the selection of policy pilot areas is deliberately controlled, resulting in a degree of artificial selectivity in the grouping of experimental and control units. Given that the ECP policies examined in this study have multiple time points, the traditional Difference-in-Differences (DID) model, which is typically used to assess policy effects at a single time point, is not suitable. Therefore, this study adopted the approach proposed by Beck et al. [27] to construct a DID model with multiple periods in order to evaluate the impact of ECP policies on carbon emissions. The specific formula used is as follows:

$$EI_{it} = \alpha_i + \beta_i \cdot Post_{it} \cdot Treat_{it} + \sum \gamma \cdot Controls_{it} + v_{year} + \mu_{city} + \varepsilon_{it} \quad (1)$$

$$EFF_{it} = \alpha_i + \beta_i \cdot Post_{it} \cdot Treat_{it} + \sum \gamma \cdot Controls_{it} + v_{year} + \mu_{city} + \varepsilon_{it} \quad (2)$$

where EI_{it} is the carbon intensity in the city i and year t, EFF_{it} represents the carbon efficiency in the city i and year t. $Post_{it}$ and $Treat_{it}$ are time and policy dummy variables. Specifically, if the city i is selected for the ECP Policies, its policy dummy variable $Treat_{it} = 1$; otherwise, it is set to 0. Similarly, if the city i is included in the ECP Policies in the year t, the $Post_{it}$ and $Post_{i,t+n}$ are all set to 1; otherwise, $Post_{it} = 0$. The interaction term

$Post_{it} \times Treat_{it}$ is the explanatory variable in the model. Furthermore, we included a set of control variables denoted as $Controls_{it}$ to account for other potential factors that may influence carbon intensity and efficiency. The variables v_{year} and μ_{city} represent time-fixed effects and city-fixed effects, respectively. The error term is represented as ε_{it}.

3.2. Impact Mechanisms

Apart from direct regulatory measures, ECP policies commonly exert influence on carbon emissions through two primary pathways: technological progress and industrial optimization. Drawing upon the methodology suggested by Akerman et al. [28], we incorporated interaction terms into a DID model to discern the distinct effects between the intergroup coefficient and the policy. The formula employed is as follows:

$$EI_{it} = \alpha_1 + \beta_1 \cdot Post_{it} \cdot Treat_{it} + \beta_2 \cdot Post_{it} \cdot Treat_{it} \cdot Effect_{it} + \sum \gamma \cdot Controls_{it} + v_{year} + \mu_{city} + \varepsilon_{it} \quad (3)$$

$$EFF_{it} = \alpha_1 + \beta_1 \cdot Post_{it} \cdot Treat_{it} + \beta_2 \cdot Post_{it} \cdot Treat_{it} \cdot Effect_{it} + \sum \gamma \cdot Controls_{it} + v_{year} + \mu_{city} + \varepsilon_{it} \quad (4)$$

When examining the impact of technological progress, substitute the technology effect variable $Tech_{it}$ for $Effect_{it}$. When assessing the effect of industrial optimization, replace $Effect_{it}$ with the industrial effect variable $Oppti_{it}$, and then compare the differences between β_1 and β_2 to discern the policy's influence. The data source is the "China City Statistical Yearbook".

3.3. Incentive–Constraint Mechanisms

3.3.1. Incentive Mechanisms

China's current territorial and quantified evaluation system serves as a strong motivator for local officials. This "incentive-oriented" instrument capitalizes on the competitive nature of promotions among local officials, functioning as a mechanism to stimulate their dedication. Importantly, it fosters a spirit of mutual competition among local officials, thereby further incentivizing local governments to intensify their efforts in implementing ECP policies, all in the pursuit of their own self-interest.

We constructed incentive variables based on the important indicators of local officials' promotion tournament, the incentive virtual variable Int_{it} and $Post_{it} \times Treat_{it}$, to construct a triple interaction term. Based on relevant studies [29], empirical testing is conducted using the DDD model with the following formulas:

$$EI_{it} = \alpha_1 + \beta_1 \cdot Post_{it} \cdot Treat_{it} \cdot Int_{it} + \sum \gamma \cdot Controls_{it} + v_{year} + \mu_{city} + \varepsilon_{it} \quad (5)$$

$$EFF_{it} = \alpha_1 + \beta_1 \cdot Post_{it} \cdot Treat_{it} \cdot Int_{it} + \sum \gamma \cdot Controls_{it} + v_{year} + \mu_{city} + \varepsilon_{it} \quad (6)$$

Regarding the Int_{it} variable, we stipulate that if the mayor or party secretary of a city is promoted in the current year, the official promotion index for that year is defined as 1. If both the mayor and the party secretary are promoted, the concept's official promotion index is defined as 2. The cumulative intensity of official promotion within 12 years is then calculated to determine the official promotion frequency of the city. The data on official changes in Chinese cities were collected manually by the author through the internet.

3.3.2. Constraint Mechanisms

Technological innovation is a critical factor in the effectiveness of ECP policies in promoting the reduction of carbon emissions. In the context of vertical constraints between central and local governments, local governments are likely to use the number of patents as a performance indicator for regional innovation when reporting to the central government in order to serve their self-interest. Therefore, we decomposed the dependent variable into "substantial innovation" and "strategic innovation", constructing a DID model to examine the constraint intensity of different policies, with the following formulas:

$$TI_{it} = \alpha_1 + \beta_1 \cdot Post_{it} \cdot Treat_{it} + \sum \gamma \cdot Controls_{it} + v_{year} + \mu_{city} + \varepsilon_{it} \tag{7}$$

$$SI_{it} = \alpha_1 + \beta_1 \cdot Post_{it} \cdot Treat_{it} + \sum \gamma \cdot Controls_{it} + v_{year} + \mu_{city} + \varepsilon_{it} \tag{8}$$

TI_{it} represents "substantive innovation", using the number of green patent inventions as a proxy variable while SI_{it} represents "strategic innovation", using the number of utility model patents and design patents as proxy variables. The patent data comes from the China State Intellectual Property Office.

3.4. Dependent Variable

To measure the carbon emission reduction effects, this paper assesses the carbon emissions of various prefecture-level cities in China from two dimensions: carbon intensity and carbon efficiency.

Carbon intensity (EI_{it}) refers to the carbon dioxide emissions per unit of GDP, reflecting the relative relationship between economic growth and carbon emissions. The calculation of carbon dioxide emissions involves aggregating the carbon emissions resulting from coal gas and liquefied petroleum gas consumption, electricity usage, transportation activities, and heat energy consumption within each city. The carbon emissions resulting from the consumption of coal gas and liquefied petroleum gas are calculated using conversion factors provided by the Intergovernmental Panel on Climate Change (IPCC) in 2006. The carbon emissions resulting from electricity consumption are calculated using the emission factors of the regional power grid [30]. The carbon emissions generated by transportation are calculated using the passenger and freight volumes of different transportation modes in the city [31]. Lastly, the carbon emissions resulting from thermal energy consumption are calculated by considering the amount of raw coal consumed by the city. All carbon emissions are then added together to obtain the total carbon emissions.

Carbon efficiency (EFF_{it}) is measured using the slacks-based model (SBM) with non-desirable outputs proposed by Tone [32]. In this model, Chinese cities are considered as distinct decision-making units, each having three vectors: inputs, expected outputs, and unexpected outputs. Specifically, the input vector can be denoted as $X \in R^m$, the expected output vector as $Y \in R^q$, and the non-expected output vector as $B \in R^p$. The input matrix, expected output matrix, and non-expected output matrix can be defined as follows: $X = [x_1,x_2...,x_n] = R^{(m\times n)}$, $Y = [y_1,y_2...,y_n] = Y^{(q\times n)}$, $B = [b_1,b_2...,b_n] = R^{(p\times n)}$.

Assuming $X > 0$, $Y > 0$, $B > 0$, the production possibility set is $P = \{(x,y,b) \mid x \geq X_\lambda,\ y \geq Y_\lambda,\ b \geq B_\lambda,\ \lambda > 0\}$, λ is weight vector, ρ^* is the objective function.

$$\rho^* = min \frac{1 - \frac{1}{m}\sum_{i=1}^{m} \frac{s_i^-}{x_{ik}}}{1 + \frac{1}{k_1+k_2}\left(\sum_{r=1}^{k_1} \frac{s_r^+}{y_{rk}} + \sum_{t=1}^{k_2} \frac{s_t^-}{b_{tk}}\right)} \tag{9}$$

$$\begin{aligned} s.t.\ x_{ik}| &= \sum_{j=1}^{n} \lambda_j x_{ij} + s_i^- \quad i = 1,2,\ldots,m; \\ y_{rk} &= \sum_{j=1}^{n} \lambda_j y_{rj} + s_r^+ \quad r = 1,2,\ldots,q; \\ b_{tk} &= \sum_{j=1}^{n} \lambda_j b_{tj} + s_t^- \quad t = 1,2,\ldots,p; \\ &\lambda_j \geq 0,\ s_i^- \geq 0,\ s_r^+ \geq 0,\ s_t^- \geq 0 \end{aligned}$$

s represents the slack variables for input vectors, expected output vectors, and non expected output vectors, while λ stands for the weight vector. When $0 < \rho^* < 1$, it indicates the presence of some efficiency loss, which means that carbon efficiency can be improved through enhancements in input and output units. When the objective function $\rho^* \geq 1$, it signifies high input–output efficiency for the region, with higher ρ^* values indicating higher carbon efficiency in that area. We use year-end employment figures as human input, the capital stock formed by fixed asset investments based on the 2006 as capital input, urban electricity generation as energy input, GDP as economic output, and carbon dioxide emissions as non-expected output to measure carbon efficiency in Chinese cities. The data used in the calculation all comes from the "China City Statistical Yearbook" and "China Energy Statistical Yearbook".

3.5. Control Variables

In view of the available statistical data, we conducted empirical testing using a panel dataset covering 224 cities in China from 2009 to 2020. Acknowledging that other urban characteristics might potentially impact carbon emissions, we included the following control variables: urbanization level, denoted as the ratio of urban population to the total regional population; financial level, quantified as the ratio of local loans to GDP; industrial level, captured by the share of value added by the secondary industry in the total output; infrastructure development level, measured as the proportion of road mileage to urban area for each prefecture-level city; fiscal decentralization level, represented by the ratio of city fiscal revenue to GDP. Data sources include the "China City Statistical Yearbook" and the National Bureau of Statistics of China.

4. Results

4.1. Main Results

The results in Table 2 show that, irrespective of the implementation of one or both policies, cities within the pilot scope have passed the test. This implies that the implementation of the ECP policies in these regions has effectively led to a reduction in carbon intensity within the respective areas. This substantiates that the ECP policies currently enforced in China have significantly promoted carbon emission reduction and positively influenced the ecological environment development of the energy industry. Furthermore, as one progresses from Policy A2 to Policy A1 and subsequently to regions where both policies are simultaneously implemented the policy intensity escalates, yielding a corresponding increase in its impact on carbon intensity and carbon efficiency. Policy A1 demonstrates effects on carbon intensity and carbon efficiency, registering values of −2.088 and 0.132, which are notably higher than those of Policy A2. In cities that implement both policies concurrently, the effects of the pilot policy on carbon intensity and carbon efficiency are −2.620 and 0.182, surpassing the outcomes observed in cities implementing a single policy. This substantiates Hypothesis 1 that ECP policy can reduce the carbon intensity of pilot cities and improve carbon efficiency in the energy industry. Additionally, as the policy intensity of ECP policies gradually increases, the policy effect also improves.

In the experiment on control variables, we find that financialization, industrialization, and the enhancement of fiscal decentralization all had a significant impact in elevating carbon intensity and reducing carbon efficiency. It is noteworthy that urbanization was the sole exception, as it increased carbon intensity but did not reduce carbon efficiency. These findings align with logical reasoning and prior experimental knowledge. Interestingly, among the controlled variables, the improvement in infrastructure construction had a substantial effect on decreasing carbon intensity and enhancing carbon efficiency. While the influence of improved infrastructure development on carbon reduction is relatively modest when compared to the ECP pilot policy, it substantiates China's capacity to mitigate carbon emissions by its notable externalities.

Table 2. Main Regression.

	EI	EFF	EI	EFF	EI	EFF
A1	−2.088 ** (0.818)	0.132 *** (0.026)				
A2			−0.605 * (0.333)	−0.009 (0.011)		
A1 × A2					−2.620 ** (1.315)	0.182 *** (0.042)
Urban	2.291 ** (0.955)	−0.003 (0.031)	2.295 ** (0.956)	0.000 (0.031)	2.321 ** (0.956)	−0.005 (0.031)
Finance	1.275 *** (0.247)	−0.040 *** (0.008)	1.271 *** (0.247)	−0.039 *** (0.008)	1.272 *** (0.247)	−0.039 *** (0.008)
Industry	1.532 *** (0.321)	−0.049 *** (0.010)	1.590 *** (0.321)	−0.051 *** (0.010)	1.559 *** (0.321)	−0.050 *** (0.010)
Facility	−0.134 *** (0.044)	0.005 *** (0.001)	−0.125 *** (0.044)	0.004 *** (0.001)	−0.140 *** (0.044)	0.005 *** (0.001)
Fiscal	21.632 *** (6.291)	−1.144 *** (0.202)	22.003 *** (6.298)	−1.142 *** (0.203)	22.042 *** (6.297)	−1.173 *** (0.203)
Constant	−18.498 *** (4.428)	1.250 *** (0.142)	−19.304 *** (4.428)	1.281 *** (0.143)	−18.878 *** (4.426)	1.273 *** (0.142)
Observations	2688	2688	2688	2688	2688	2688
Adjust R^2	0.629	0.680	0.628	0.677	0.628	0.679
City fixed effects	Yes	Yes	Yes	Yes	Yes	Yes
Year fixed effects	Yes	Yes	Yes	Yes	Yes	Yes

Notes: *, **, and *** mean significant at the 10%, 5%, and 1% level, respectively.

4.2. Placebo Test

Although we controlled for urban variables in the experiment, it was imperative to enhance the robustness of our regression results. To achieve this, we randomly selected an equivalent number of cities from all sample cities as the control group for placebo tests [33]. Employing a random sampling approach, we generated 500 sets of placebo variables. The resulting kernel density and coefficient distributions were then depicted in figures, allowing for a comparison with the original findings and the presentation of the placebo test outcomes.

Figure 1 illustrates the results of the placebo tests for Policy A1 and Policy A2. Notably, the sampled outcomes of carbon intensity and carbon efficiency for Policy A1 significantly deviate from the original results, substantiating that our experimental findings do not exhibit a placebo effect. While the effect of Policy A2 on carbon intensity also passes the test, the results for carbon efficiency, due to their lack of significance within the policy itself, do not exhibit significant differences compared to the placebo test. These test results are consistent with the main regression findings, demonstrating the robustness of the experiment.

4.3. Impact Mechanisms Results

Tables 3 and 4 present how ECP policies influence carbon reduction in the urban green energy industry. In Table 3, we find after introducing the interaction terms in the DID model that the estimated value β_2 of the interaction term is significantly smaller than the estimated value β_1 of ECP policies for carbon intensity. Similar results manifest in the context of carbon efficiency. These findings substantiate the implications of energy industrial structure optimization, thus affirming Hypothesis 2a. ECP policies effectively reduce carbon intensity and enhance carbon efficiency by stimulating energy industrial structure optimization. The results in Table 4 are consistent with those in Table 3. The incorporation of interaction terms leads to a notable decrease in carbon intensity and a significant improvement in carbon efficiency, thus providing empirical support for Hypothesis 2b. ECP policies can reduce carbon intensity and enhance carbon efficiency by promoting technological progress.

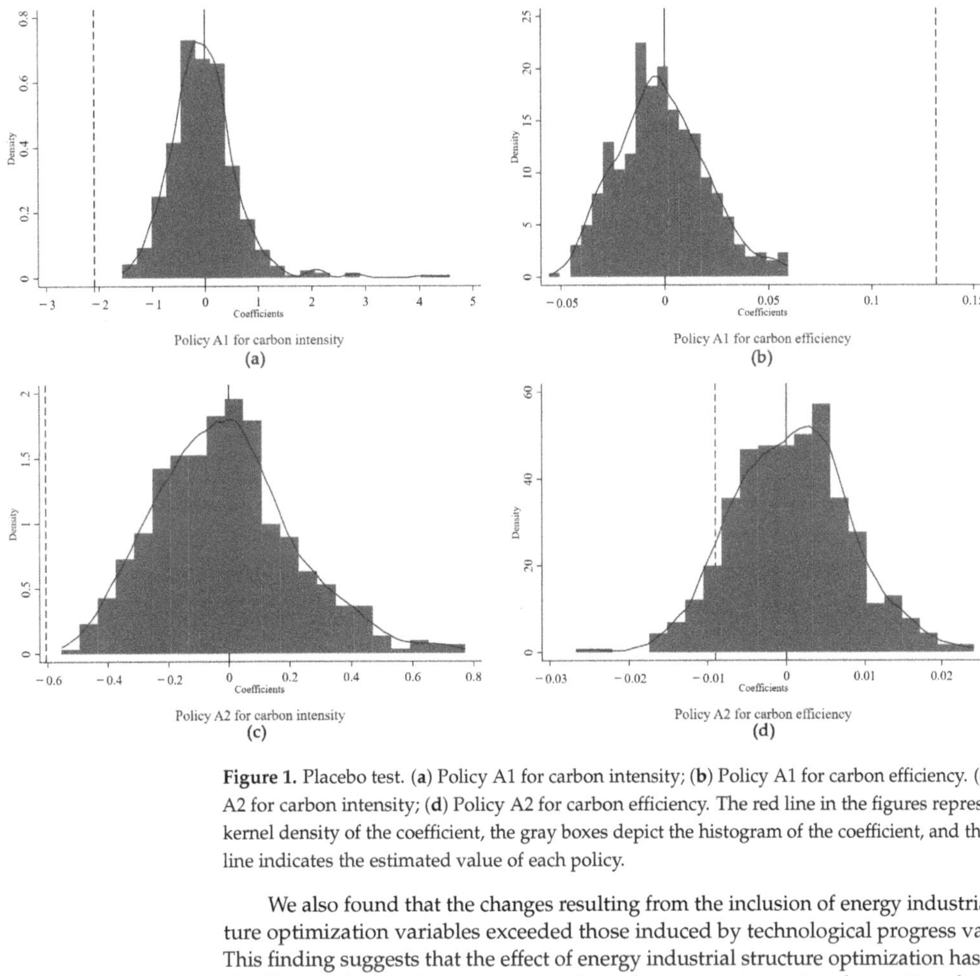

Figure 1. Placebo test. (**a**) Policy A1 for carbon intensity; (**b**) Policy A1 for carbon efficiency. (c) Policy A2 for carbon intensity; (**d**) Policy A2 for carbon efficiency. The red line in the figures represents the kernel density of the coefficient, the gray boxes depict the histogram of the coefficient, and the dotted line indicates the estimated value of each policy.

We also found that the changes resulting from the inclusion of energy industrial structure optimization variables exceeded those induced by technological progress variables. This finding suggests that the effect of energy industrial structure optimization has a more significant influence than green technological progress. As previously mentioned, whether from the Pollution Haven Theory or the Environmental Regulation Theory perspective, the optimization of industrial structure has a positive impact on carbon reduction objectives. In contrast, technological progress exhibits a dual effect, both promoting carbon reduction through "innovation compensation" and lowering carbon efficiency due to "compliance costs". Although the experiment didn't conclusively affirm this proposition, it indirectly lends credence to its plausibility.

4.4. *Incentive-Constraint Mechanism*

Local governments have historically played a significant role in fostering regional economic and social progress. Within the context of the specific central–local interaction mechanism, it is important to highlight the dual roles played by the central government. On one hand, the central government employs constraint mechanisms, which exert a vertical constraint effect on local governments. This serves as a vital means to mitigate potential moral hazards that may arise among local governments. On the other hand, the central government strategically employs incentive mechanisms to motivate local governments, acting as self-interested actors, to actively pursue elevated economic and political benefits.

Table 3. How ECP policies affect carbon emissions reductions by industrial structure optimization.

	EI	EFF	EI	EFF
A1	36.092 ** (16.210)	−1.253 ** (0.521)		
A1 × Ins	−5.595 ** (2.372)	0.203 *** (0.076)		
A2			15.959 *** (4.900)	−0.387 ** (0.158)
A2 × Ins			−2.511 *** (0.741)	0.057 ** (0.024)
Constant	−18.451 *** (4.424)	1.248 *** (0.142)	−18.886 *** (4.420)	1.272 *** (0.143)
Observations	2688	2688	2688	2688
Adjust R^2	0.630	0.681	0.630	0.678
Control	Yes	Yes	Yes	Yes
City fixed effects	Yes	Yes	Yes	Yes
Year fixed effects	Yes	Yes	Yes	Yes

Notes: ** and *** mean significant at the 5%, and 1% level respectively.

Table 4. How ECP policies affect carbon emissions reductions by technological progress.

	EI	EFF	EI	EFF
A1	5.053 * (2.620)	−0.149 * (0.084)		
A1 × Tech	−0.998 *** (0.348)	0.039 *** (0.011)		
A2			2.480 *** (0.931)	−0.193 *** (0.030)
A2 × Tech			−0.524 *** (0.148)	0.031 *** (0.005)
Constant	−18.506 *** (4.427)	1.249 *** (0.142)	−18.603 *** (4.426)	1.239 *** (0.142)
Observations	2678	2678	2678	2678
Adjust R^2	0.631	0.681	0.631	0.682
Control	Yes	Yes	Yes	Yes
City fixed effects	Yes	Yes	Yes	Yes
Year fixed effects	Yes	Yes	Yes	Yes

Notes: * and *** mean significant at the 10% and 1% level respectively.

4.4.1. Incentive Mechanism

The effectiveness of "horizontal competition" stems from the capacity of local governments, operating within this framework, to optimize their interests through the exercise of administrative authority. Consequently, for the highest-ranking executive officials within local governments, the strength of incentives, reflecting the alignment of interests among these officials, takes the form of a "promotion" system within the Chinese context. In this system, when a local government official demonstrates outstanding abilities in the region under their jurisdiction, they frequently experience expedited career advancement. This dynamic underscores one of the merits of the promotion tournament, as it facilitates the rapid elevation of competent officials.

Table 5 reveals the examination results of the incentive mechanism. Regions that have implemented ECP Policies, whether it is Policy A1 or Policy A2, exhibit significant correlation between the level of official promotions and the effectiveness of the policy. This phenomenon indicates the presence of an official promotion tournament within China's pilot policies. These ECP policies can promote carbon reduction through the mechanisms related to official promotions. In the process of implementing ECP policies, when a region has a higher intensity of official promotions, its local leaders will exert more effort to enhance the policy effects of the ecological civilization pilot, thereby positioning themselves more favorably in the official promotion tournament.

Table 5. Impact of policy incentive mechanism.

	(1)	(2)	(3)	(4)
	EI	EFF	EI	EFF
A1 × Int	−3.974 *** (1.418)	0.244 *** (0.046)		
A2 × Int			−0.982 * (0.536)	−0.004 (0.017)
Constant	−18.559 *** (4.425)	1.255 *** (0.142)	−19.289 *** (4.427)	1.284 *** (0.143)
Observations	2688	2688	2688	2688
Adjust R^2	0.629	0.680	0.628	0.677
Control	Yes	Yes	Yes	Yes
City fixed effects	Yes	Yes	Yes	Yes
Year fixed effects	Yes	Yes	Yes	Yes

Notes: * and *** mean significant at the 10% and 1% level, respectively.

In this study, we also observed that, in comparison to areas which implemented Policy A2, Policy A1 demonstrated a more pronounced promotion of carbon reduction and enhancement of carbon efficiency. This observation indicates that Policy A1 as a "green honor", can effectively motivate local officials to elevate their ecological civilization development efforts. Such endeavors result in decreased carbon emissions and improved carbon efficiency. Contrarily, Policy A2, which offers long-term incentives, failed to produce a similar effect on carbon efficiency through the official promotion tournament. These findings provide empirical support for Hypothesis 3a, indicating that ECP policies with short-term incentives are more effective in reducing carbon intensity and enhancing carbon efficiency when employing a well-structured incentive mechanism.

4.4.2. Constraint Mechanism

Table 6 presents the examination results of the constraint mechanism. Policy A2, which features an elimination system, is expected to have a stricter assessment compared to Policy A1. However, in practical policy implementation, Policy A1 follows fixed assessment criteria, while Policy A2 allows self-declared criteria. To facilitate their assessment approval, local officials tend to opt for easily achievable indicators. This tendency is particularly pronounced in terms of innovation. We find that both ECP policies have a significantly negative impact on strategic innovation. The implementation of these policies effectively reduces the level of strategic innovation. Additionally, the impact of Policy A2 on substantive innovation is notably lower than that of Policy A1. Policy A1 exhibits a significantly more substantial influence in diminishing strategic innovation when contrasted with Policy A2. This observation confirms Hypothesis 3b, suggesting that as the strength of ECP policies shifts from weak to strong, local governments tend to lean more towards strategic innovation rather than substantive innovation. This indicates that the assessment mechanism set in ECP policies makes local governments refrain from using strategic innovation to bypass central governments. The increased constraint does indeed somewhat reduce the behavior of local governments pursuing innovation quantity over innovation quality for political gains. Importantly, the stringency of the assessment is not the sole factor; the reasonable design of assessment criteria has the potential to more effectively promote substantive innovation while diminishing strategic innovation.

Table 6. Impact of policy constraint mechanism.

	Substantive Innovation	Strategic Innovation	Substantive Innovation	Strategic Innovation
A1	0.230 * (0.131)	−0.282 ** (0.112)		
A2			−0.149 *** (0.053)	−0.151 *** (0.045)
Constant	2.922 *** (0.707)	1.003 * (0.605)	2.922 *** (0.706)	0.866 (0.604)
Observations	2688	2688	2688	2688
Adjust R^2	0.901	0.925	0.901	0.925
Control	Yes	Yes	Yes	Yes
City fixed effects	Yes	Yes	Yes	Yes
Year fixed effects	Yes	Yes	Yes	Yes

Notes: *, **, and *** mean significant at the 10%, 5%, and 1% level respectively.

Combined with the results in Tables 5 and 6, it is evident that the incentive–constraint mechanism formulated by Policy A1 is more reasonable. Both its incentive and constraint mechanisms contribute to the enhancement of regional ecological civilization construction, thereby promoting the reduction of carbon emissions and the enhancement of carbon efficiency within the region. In contrast, Policy A2 leads to a concurrent decline in both strategic and substantial innovation levels. In regions where ECP policies are implemented with heightened assessment intensity, the continuous rise in "compliance costs" results in a crowding-out effect, leading to a decrease in the level of "substantial innovation" among enterprises. Concurrently, due to the formidable constraint capacity of the central government, regions that implement ECP policies witness a decrease in the level of "strategic innovation" and it is evident that the incentive–constraint mechanism formulated by Policy A1 is more reasonable. Both its incentive and constraint mechanisms contribute to the enhancement of regional ecological civilization construction, thereby promoting the reduction of carbon emissions and the enhancement of carbon efficiency within the region. In contrast, Policy A2, characterized by higher assessment intensity, leads to a concurrent decline in both strategic and substantial innovation levels. In regions where ECP policies are implemented with heightened assessment intensity, the continuous rise in "compliance costs" results in a crowding-out effect, leading to a decrease in the level of "substantial innovation" among enterprises. Concurrently, due to the formidable constraint capacity of the central government, regions that implement ECP policies witness a decrease in the level of "strategic innovation".

5. Conclusions

In this study, we have empirically verified Hypotheses 1 through 3, leading us to the unequivocal conclusion that China's current ECP policies are highly effective in reducing regional carbon emissions. This reduction is principally accomplished through the optimization of industrial structures and the advancement of technological capabilities. Additionally, the Chinese incentive–constraint mechanism plays a pivotal role in this endeavor. Short-term incentives and well-defined assessment standards serve as motivational levers, prompting active participation among local officials in ecological civilization development. These findings not only provide empirical support for the theoretical foundation of ECP policies but also present practical strategies for establishing administrative frameworks dedicated to fostering carbon emission reduction.

Furthermore, our study highlights the evolving role of the official promotion competition in China's governance framework. It no longer solely prioritizes GDP growth but has progressively encompassed ecological civilization goals. This transformation endows it with substantial influence in incentivizing local government officials to contribute proactively to environmental and ecological initiatives. Our research also uncovers a positive correlation between the comprehensiveness of evaluation criteria within ECP policies and their actual policy impacts. In contrast, permitting local governments to independently formulate assessment standards, while considering local contexts to some extent, does not inherently encourage substantial innovation at the grassroots level. This underscores the critical importance of judicious central government oversight applying reasonable constraints on their subordinate counterparts, thereby fostering greater dedication to ecological civilization construction and ultimately enhancing policy effectiveness.

Based on these findings, we offer the following three recommendations for the central government:

Global Adoption of ECP Policies for Carbon Emission Reduction. It is recommended that governments worldwide consider implementing policies similar to China's ECP policies. These policies have demonstrated their effectiveness in reducing regional carbon emissions. By adopting ECP-like initiatives, governments can advance their own carbon emission reduction efforts. Emulating successful models and adapting them to local contexts can provide a structured framework for addressing carbon reduction and environmental sustainability at the national level.

Enhanced Guidance for the Energy Industry and Promoting Technological Innovation. To expedite carbon emission reduction and bolster ecological objectives, governments should provide heightened guidance to the energy industry. Policymakers can encourage technological innovation within the sector, emphasizing the development of environmentally friendly and efficient technologies. This approach not only supports carbon reduction but also fuels economic growth by fostering technological advancements that align with ecological and sustainability goals.

Tailored Incentive and Constraint Strategies in ECP Policy Design. When crafting ECP policies, policymakers should take into account the incentive and constraint effects inherent in the policy design. It is essential to customize these strategies based on the specific context and conditions of each country. This tailored approach ensures that the incentive and constraint mechanisms are well-suited to the unique circumstances of each nation. By doing so, governments can maximize the effectiveness of their ecological policies and motivate active participation among stakeholders.

However, it is important to acknowledge the limitations of this research. We focused primarily on China and did not consider the effects of environmental development policies in other nations. In future research, we plan to integrate data from other countries to broaden our understanding of the efficacy of environmental policies on a global scale. Additionally, while we have conducted robustness tests to address endogeneity, we aim to explore more rigorous methods for handling endogeneity and reducing potential estimation biases in later stages. This is in line with the advancement of quantitative research techniques and will contribute to a more comprehensive understanding of the subject.

In summary, our study contributes to the understanding of the impact of ECP policies on carbon emissions in the urban green energy industry. The policy recommendations presented here aim to further enhance environmental policies and carbon reduction efforts in China, while recognizing the need for more extensive research and methodological advancements in the field.

Author Contributions: Conceptualization, P.Z. and F.L.; Data curation, L.T.; Methodology, P.Z. and F.L.; Resources, F.L.; Software, L.T.; Supervision, F.L.; Writing—original draft, P.Z. and L.T.; Writing—review & editing, P.Z., L.T. and F.L. All authors have read and agreed to the published version of the manuscript.

Funding: This research received no external funding.

Data Availability Statement: The data that support the findings of this study are available from the corresponding author upon request.

Conflicts of Interest: The authors declare no conflict of interest.

References

1. Heurtebise, J.Y. Sustainability and ecological civilization in the age of Anthropocene: An epistemological analysis of the psychosocial and "culturalist" interpretations of global environmental risks. *Sustainability* **2017**, *9*, 1331. [CrossRef]
2. Pickering, J.; Bäckstrand, K.; Schlosberg, D. Between environmental and ecological democracy: Theory and practice at the democracy-environment nexus. *J. Environ. Policy Plan.* **2020**, *22*, 1–15. [CrossRef]
3. Hansen, M.H.; Li, H.; Svarverud, R. Ecological civilization: Interpreting the Chinese past, projecting the global future. *Glob. Environ. Chang.* **2018**, *53*, 195–203. [CrossRef]
4. Zhang, Y.; Fu, B. Impact of China's establishment of ecological civilization pilot zones on carbon dioxide emissions. *J. Environ. Manag.* **2023**, *325*, 116652. [CrossRef]
5. Van Leeuwen, G.; Mohnen, P. Revisiting the Porter hypothesis: An empirical analysis of green innovation for the Netherlands. *Econ. Innov. New Technol.* **2017**, *26*, 63–77. [CrossRef]
6. Porter, M.E.; van der Linde, C. Toward a new conception of the environment-competitiveness relationship. *J. Econ. Perspect.* **1995**, *9*, 97–118. [CrossRef]
7. Wang, X.; Zhang, T.; Nathwani, J.; Yang, F.; Shao, Q. Environmental regulation, technology innovation, and low carbon development: Revisiting the EKC Hypothesis, Porter Hypothesis, and Jevons' Paradox in China's iron & steel industry. *Technol. Forecast. Soc. Chang.* **2022**, *176*, 121471. [CrossRef]
8. Gabel, H.L.; Sinclair-Desgagné, B. Managerial incentives and environmental compliance. *J. Environ. Econ. Manag.* **1993**, *24*, 229–240. [CrossRef]
9. Wu, J.; Deng, Y.; Huang, J.; Morck, R.; Yeung, B. *Incentives and Outcomes: China's Environmental Policy*; National Bureau of Economic Research: Cambridge, MA, USA, 2013. [CrossRef]
10. Zhang, J.; Liu, Y.; Chang, Y.; Zhang, L. Industrial eco-efficiency in China: A provincial quantification using three-stage data envelopment analysis. *J. Clean. Prod.* **2017**, *143*, 238–249. [CrossRef]
11. Hu, J.; Wu, Y.; Irfan, M.; Hu, M. Has the ecological civilization pilot promoted the transformation of industrial structure in China? *Ecol. Indic.* **2023**, *155*, 111053. [CrossRef]
12. Ercolano, S.; Romano, O. Spending for the environment: General government expenditure trends in Europe. *Soc. Indic. Res.* **2018**, *138*, 1145–1169. [CrossRef]
13. Zhang, L.; Wang, H.; Zhang, W.; Wang, C.; Bao, M.; Liang, T.; Liu, K. Study on the development patterns of ecological civilization construction in China: An empirical analysis of 324 prefectural cities. *J. Clean. Prod.* **2022**, *367*, 132975. [CrossRef]
14. Copeland, B.R.; Taylor, M.S. Trade, growth, and the environment. *J. Econ. Lit.* **2004**, *42*, 7–71. [CrossRef]
15. Ren, Y.S.; Boubaker, S.; Liu, P.Z.; Weber, O. How does carbon regulatory policy affect debt financing costs? Empirical evidence from China. *Q. Rev. Econ. Financ.* **2023**, *90*, 77–90. [CrossRef]
16. Buccella, D.; Fanti, L.; Gori, L. To abate, or not to abate? A strategic approach on green production in Cournot and Bertrand duopolies. *Energy Econ.* **2021**, *96*, 105164. [CrossRef]
17. Zhao, T.; Zhou, H.; Jiang, J.; Yan, W. Impact of green finance and environmental regulations on the green innovation efficiency in China. *Sustainability* **2022**, *14*, 3206. [CrossRef]
18. Bai, D.; Hu, J.; Irfan, M.; Hu, M. Unleashing the impact of ecological civilization pilot policies on green technology innovation: Evidence from a novel SC-DID model. *Energy Econ.* **2023**, *125*, 106813. [CrossRef]
19. Paşcalău, E.M. The Competences of Public Environmental Protection Authorities in Case of Natural Disasters. *Perspect. Law Public Adm.* **2022**, *11*, 111–124.
20. Peráček, T.; Majerčáková, D.; Mittelman, A. Significance of the waste act in the context of the right to protection of the environment. In Proceedings of the SGEM 2016 Conference, Albena, Bulgaria, 30 June–6 July 2016; Volume 1, pp. 979–986.
21. Panigaj, J.; Bernikova, E. Ecocide—A new crime under international law? *Jurid. Trib.* **2023**, *13*, 5–20. [CrossRef]
22. Zhou, L.A. The administrative subcontract: Significance, relevance and implications for intergovernmental relations in China. *Chin. J. Sociol.* **2016**, *2*, 34–74. [CrossRef]
23. Kunrong, S.; Gang, J. The policy effects of the environmental governance of Chinese local governments: A study based on the progress of the River Chief System. *Soc. Sci. China* **2020**, *41*, 87–105. [CrossRef]
24. Ran, R. Perverse incentive structure and policy implementation gap in China's local environmental politics. *J. Environ. Policy Plan.* **2013**, *15*, 17–39. [CrossRef]

25. Kostka, G.; Nahm, J. Central–local relations: Recentralization and environmental governance in China. *China Q.* **2017**, *231*, 567–582. [CrossRef]
26. Comino, S.; Graziano, C. How many patents does it take to signal innovation quality? *Int. J. Ind. Organ.* **2015**, *43*, 66–79. [CrossRef]
27. Beck, T.; Levine, R.; Levkov, A. Big bad banks? The winners and losers from bank deregulation in the United States. *J. Financ.* **2010**, *65*, 1637–1667. [CrossRef]
28. Akerman, A.; Leuven, E.; Mogstad, M. Information frictions, internet, and the relationship between distance and trade. *Am. Econ. J. Appl. Econ.* **2022**, *14*, 133–163. [CrossRef]
29. Kang, C.; Chen, Z.; Zhang, H. The outgoing audit of natural resources assets and enterprise productivity: New evidence from difference-in-differences-in-differences in China. *J. Environ. Manag.* **2023**, *328*, 116988. [CrossRef]
30. Glaeser, E.L.; Kahn, M.E. The greenness of cities: Carbon dioxide emissions and urban development. *J. Urban Econ.* **2010**, *67*, 404–418. [CrossRef]
31. Li, H.; Lu, Y.; Zhang, J.; Wang, T. Trends in road freight transportation carbon dioxide emissions and policies in China. *Energy Policy* **2013**, *57*, 99–106. [CrossRef]
32. Tone, K. A slacks-based measure of efficiency in data envelopment analysis. *Eur. J. Oper. Res.* **2001**, *130*, 498–509. [CrossRef]
33. Liu, Q.; Lu, Y. Firm investment and exporting: Evidence from China's value-added tax reform. *J. Int. Econ.* **2015**, *97*, 392–403. [CrossRef]

energies

MDPI

Article

New Energy Commuting Optimization under Low-Carbon Orientation: A Case Study of Xi'an Metropolitan Area

Xin Dai [1,*], Tianshan Ma [2] and Enyi Zhou [3,*]

1 School of Transportation Engineering, Chang'an University, Xi'an 710061, China
2 School of Economics and Management, Chang'an University, Xi'an 710061, China; mtshan@chd.edu.cn
3 School of Economics and Management, Xi'an University of Architecture and Technology Huaqing College, Xi'an 710043, China
* Correspondence: dxin0914@163.com (X.D.); 2019023006@chd.edu.cn (E.Z.)

Abstract: Low-carbon travel is an important part of low-carbon cities and low-carbon transportation, and low-carbon transportation is an inevitable choice to slow down the growth of carbon emissions in China. All countries in the world are actively promoting new energy vehicles and attach great importance to the application of the new energy industry in urban transportation. Commuting is an important part of urban life, and the choice of travel behavior has an important impact on traffic and environmental protection. Taking the Xi'an metropolitan area as an example, this paper expounds on the integrated development path of the industrial chain of new energy + travel in the metropolitan area and clarifies the energy transformation model of the integrated development of low-carbon transportation and energy. From the perspective of green and low-carbon, 1000 commuters were interviewed using a questionnaire survey, and the cumulative prospect model was used to verify the internal mechanism affecting commuters in metropolitan areas to choose new energy commuting. The results of the study show that new energy transportation modes play an important role in the low-carbon economy, and under different scenarios and assumptions, there are significant differences in the cumulative prospect values of the subway, new energy buses and fuel private cars, and corresponding optimization measures are proposed to increase the proportion of new energy commuting trips. The results will help further promote the development of a low-carbon economy and energy integration in the field of transportation and provide a reference for the sustainable development of public transportation.

Keywords: low-carbon; green travel; metropolitan area; new energy; commuting

Citation: Dai, X.; Ma, T.; Zhou, E. New Energy Commuting Optimization under Low-Carbon Orientation: A Case Study of Xi'an Metropolitan Area. *Energies* **2023**, *16*, 7916. https://doi.org/10.3390/en16237916

Academic Editor: Seung-Hoon Yoo

Received: 11 November 2023
Revised: 28 November 2023
Accepted: 2 December 2023
Published: 4 December 2023

1. Introduction

With the continuous acceleration of economic development and urbanization, the proportion of transportation in national energy consumption and carbon emissions is rising, and it will face increasingly severe resource and environmental constraints. Sustainable development is one of the most pressing challenges facing mankind in the 21st century, among which energy consumption and climate warming have become the focus of global attention and research hotpot, and actively responding to climate change and promoting green and low-carbon development is a key link [1]. A metropolitan area is a new urban regional form with symbolic significance in the process of modern social and economic development and is an inevitable trend of urbanization development in countries around the world. It has become the core area of China's economic development and carbon dioxide emissions. The transportation system has always been one of the main ways to achieve trans-regional large-scale transportation of energy, which profoundly affects the layout of China's energy production. In addition, the accelerated expansion of the transport system and the trend towards re-electrification further strengthen the link between the transport system and the energy system, becoming a key factor affecting the efficiency of the energy system operation [2].

On the one hand, the whole energy industry chain is facing a profound impact. According to the data, the terminal energy consumption of China's transportation sector in 2022 is 820 million tons of standard coal. The total carbon emissions from the transportation industry in Shaanxi Province from 2010 to 2022 were 3.089 million tons. The main types of energy consumption in Shaanxi are gasoline and diesel, which are the main sources of carbon emissions from transportation. The intensity of transportation plays an important role in suppressing the growth of carbon emissions. Specifically, the contribution value of the transportation intensity factor is negative at 2.1724 million tons, with a contribution rate of 0.9601 [3]. The opportunities and challenges brought by "carbon peak" and "carbon neutrality" to the transportation field not only lie in the transportation itself but also penetrate into different business links in the whole industry chain, including not only transportation equipment manufacturing, aviation, railway, road and shipping and other transportation segments, but also sales, transportation, and other service industries [4]. The industrial chain not only needs to pay attention to the proportion of renewable energy, such as non-fossil energy, available in the whole country but also extends from the structure of the energy supply source to the diversity of subsequent energy consumption, which will be affected by low-carbon development ideas. Specifically, the dimensions of the impact of a low-carbon economy on related industries in the field of transportation include differences in the means of transport itself and differences in modes of transport [5]. The former emphasizes the use of a variety of means of transport with a variety of energy structures, such as electric vehicles, diesel vehicles, gasoline vehicles, hydrogen vehicles, natural gas vehicles, etc., while the latter involves the choice of different modes of transport such as ports, railways, highways, and aviation. Therefore, in the field of transportation, to achieve the strategic task goal of "double carbon" as soon as possible, it is necessary to seek feasible solutions from multiple dimensions and implement them around the whole industry chain and the whole process to achieve good results.

On the other hand, transportation commuting, as a kind of generative demand, is the periodic and regular travel behavior of people to and from the workplace and residence. China has entered the metropolitan era of urbanization and quality improvement. The renewal and upgrading of regional spatial structure has also brought new pressure to commuting, with the rapid growth of car ownership and the continuous growth of carbon emissions in the transportation sector [6]. In recent years, the rapid economic and social development of our country has vigorously promoted the development process of urbanization and motorization, and the rapid growth of motor vehicle ownership has become an inevitable trend of social development. The heavy use of motor vehicles is one of the main reasons for the continuous increase in carbon dioxide emissions [7]. At the same time, the pursuit of beautiful and convenient travel demand by urban residents makes the number of motor vehicles continue to increase, the saturation of urban roads is getting larger and larger, and overall traffic congestion has become a common problem in major cities. The traffic carrying capacity of the inner core circle of the metropolitan area obviously exceeds the load, and the efficiency of traffic management is low, which eventually leads to the spread of traffic congestion in a larger area. It has intensified comprehensive problems such as urban environmental pollution.

Based on this, at present, the academic community focuses more on the research vision of the personal will of new energy travel and the development and promotion of new energy. Less attention is paid to the optimization of new energy in commuting. Commuting is an important part of urban life, and the choice of travel behavior has an important impact on traffic and environmental protection. Therefore, this paper takes the Xi'an metropolitan area as an example, summarizes the integration mechanism of the new energy + travel industry chain in the metropolitan area, and clarifies the energy transformation model of low-carbon transportation and energy integration development. From the perspective of green and low carbon, the cumulative prospect model was used to verify the internal mechanism affecting commuters in metropolitan areas to choose new energy commuting.

It is expected that the research results will contribute to expanding the proportion of new energy travel in metropolitan areas in the future.

2. Literature Review

The concept of low-carbon was clearly put forward after 2000; however, the idea of low-carbon travel is not a new concept; it has experienced a long-term evolution, development, and heat process. The transition from private to public transport systems is analyzed, and it is suggested that public transport systems can reduce energy demand, carbon emissions, and air pollutants in local towns. Dällenbach [8] uses cost–benefit analysis to find that a particularly effective strategy to minimize CO_2 emissions from transportation is to replace flights with rail transit, with the same train emitting about 80–90% less CO_2 than an airplane. Fletcher [9] validated that expected travel patterns also have the potential to lock in high-carbon transport and undermine progress by collecting data using an international online survey. Achieving a low-carbon mobility transition must be supported by coordinated efforts by governments and individuals. Shie [10] adopted Porter's diamond model theory to demonstrate that green commitment has a positive impact on low-carbon travel motivation and intention while it has a negative impact on low-carbon travel constraints. Liao [11] used an extended TPB model to investigate the determinants of urban residents' low-carbon travel intentions and found that attitudes, subjective norms, and perceived behavioral control have a positive impact on low-carbon travel intentions. Moriarty [12] proposed to reduce urban vehicle travel by using MSD data to analyze four methods: changing urban land use, reducing the convenience of private car travel, introducing a carbon tax, and using information technology as a travel substitute.

Some scholars' policy studies on traffic governance in the context of metropolitan areas mainly focus on the policy formulation of traffic planning and the development of public transportation, etc., and pay less attention to guiding the change of travel behavior from the level of individual commuters, so as to improve the travel structure and realize the optimization and upgrading of low-carbon traffic environment. P Næss [13], taking Norway as an example, found that reducing travel distances, promoting better transport provision, and imposing tolls on urban roads could effectively save land and reduce car travel. Abdul [14] believes that transforming traditional gasoline vehicles into new energy vehicles is an important measure to achieve low-carbon urban development goals via energy conservation and emission reduction. Electric vehicles, due to their advantages in energy conservation and carbon reduction, will play an important role in this transformation. Broin [15] limiting infrastructure deployment as a complementary policy to carbon pricing reduces the cost of mitigation.

Based on this, this paper improves the cumulative prospect theory model to explore the internal selection mechanism and application scenarios of new energy commuting travel mode selection in metropolitan areas and provides targeted countermeasures and suggestions to guide commuters to choose low-carbon travel and promote the low-carbon development of transportation organizations in metropolitan areas.

3. Integrated Development of Industrial Chain of New Energy and travel in Metropolitan Area

3.1. Relationship between Transportation Energy Consumption and Carbon Emission in Metropolitan Area

The metropolitan traffic environment system is a complex system, and the factors of the system affect and interact with each other. Dual city life, that is, the separation of the place of residence and work, has become a common phenomenon, and traffic congestion and traffic jams in the morning and evening peak have become the norm, which occupies a lot of commuters' living and working time and increases economic costs, which greatly affects the quality of life. The existing travel facilities have been unable to meet people's travel needs [16]. Therefore, the travel structure of residents has changed accordingly. The explosive growth of family cars has not only brought serious congestion of trunk roads and urban traffic but also greatly increased the consumption of oil and other harmful gases and greenhouse gas emissions [17]. The resulting energy shortage, environmental pollution, and deterioration of urban road conditions will restrict economic development and prompt the government to make policy adjustments, control private car travel, raise emission standards, and vigorously develop public transportation, thus affecting people's choice of travel modes, which will, in turn, affect energy consumption, pollution emission, and road construction, forming a complex feedback system [18]. The relationship between transportation energy consumption and carbon emissions in metropolitan areas is shown in Figure 1.

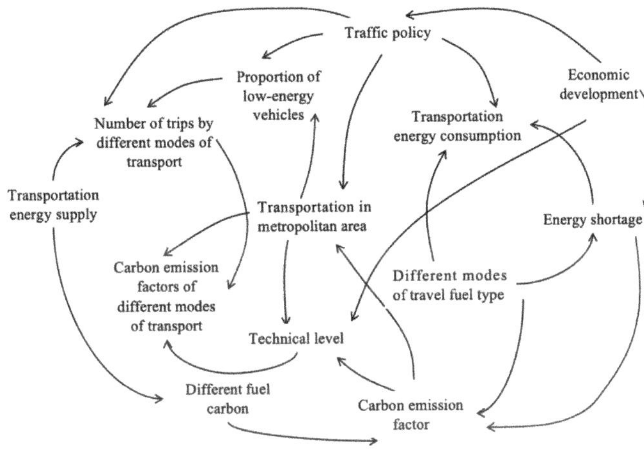

Figure 1. The relationship between transportation energy consumption and carbon emission in metropolitan areas.

3.2. Energy Integration Development of Low-Carbon Transportation

Under the background of energy Internet, the integrated energy and transportation system takes multi-network complementary as the core concept fully integrates the transportation system and deepens its development based on a multi-energy network. To realize the coordinated operation and development of energy systems such as electricity, natural gas, and heat with the railway, wheel transport, electric vehicles, urban electrified rail transit, large-scale hubs, and other transportation systems [19]. The transportation system consumes energy during transportation, so energy consumption is an important attribute attached to the basic attribute. From the energy supply side, re-electrification refers to "electricity as the center", promoting the transformation of the source of electricity from coal power generation to renewable energy power generation to solve the pollution problem in the process of energy production. By optimizing the power supply structure, we should vigorously implement clean energy substitution and electric energy substitution. From the perspective of energy consumption, re-electrification refers to "taking electricity as a priority", increasing the proportion of electric energy in terminal energy consumption, promoting efficient and clean energy utilization, and aiming to solve the problems of pollution and inefficiency in the process of energy consumption [20]. At present, the volume of "replacing oil with electricity" in transportation energy use is still small, but it is developing rapidly. In the future, through the development of electrified transportation, it can achieve "electricity instead of oil", reduce the proportion of oil in the structure of energy consumption, and slow down the growth of oil demand. At the same time, as a green energy storage carrier in the use of new energy vehicles, it is not only the main body of electricity consumption but also the main body of power supply [21]. New energy vehicles can not only reduce exhaust emissions, but intelligent shared electric vehicles can also solve the problem of travel congestion and inefficiency [22]. The energy conversion diagram for the development of low-carbon transport is shown in Figure 2.

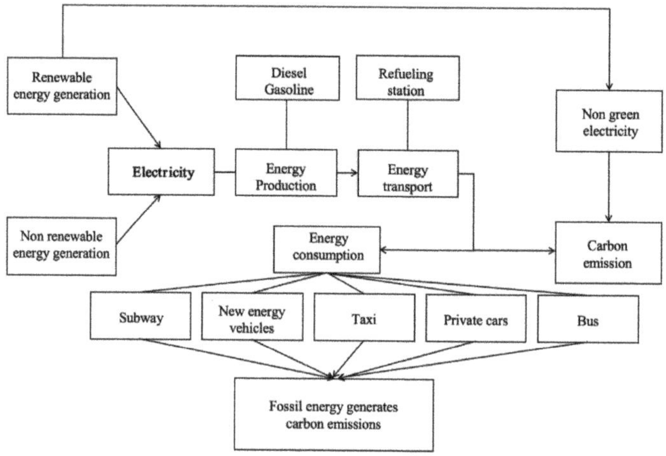

Figure 2. Energy conversion for low-carbon transport development.

3.3. Calculation Model of Average Carbon Emission of Public Trams in Metropolitan Area

In the selection of the transportation carbon emission accounting model, carbon emission is calculated according to different types of vehicle ownership, mileage, combustion per unit mileage, and combustion carbon emission coefficient [23]. The formula is shown as follows:

$$E = \sum_{m,n} Distance_{m,n} \times Consumption_{m,n} \times Density_{m} \times Calorific_{m} \times Emission\ coefficien_{m,n} \quad (1)$$

where E represents the total carbon emission of metropolitan traffic in a certain period of time; m represents the type of fuel consumed by transportation in the metropolitan area, including diesel, gasoline, natural gas, etc.; n represents the type of vehicle used for transportation in the metropolitan area. $Distance_{m,n}$ is the distance traveled by the n-type car using m fuel; $Consumption_{m,n}$ is the unit energy consumption of n vehicle using class m fuel; $Density_{m}$ is the fuel density of m fuel; $Calorific_{m}$ is the net calorific value of class m fuel; $Emission\ coefficien_{m,n}$ is the carbon emission factor of m fuel. According to this calculation, the average carbon emissions of public trams in metropolitan areas with different fuel types (Table 1) and the per capita energy consumption and carbon emission factors of individual travel modes (Table 2) are obtained.

Table 1. Average carbon emissions of public trams in metropolitan areas with different fuel types [24].

Vehicle Fuel Type	Direct Carbon Emissions (Tons)	Indirect Carbon Emissions (Tons)
Diesel oil	25	29
Natural gas	32	40
Hybrid power	22	27
Gasoline	27	33
Pure electric	0	24

Table 2. Per capita energy consumption and carbon emission factors of each mode of transportation [25].

Item	Walk	Bicycle	Bus	Subway	Taxi	Car
Per capita energy consumption (kg)	/	/	0.47	0.12	9.48	8.52
Carbon emission factor (gCO_2/kJ)	/	/	19.8	7.5	140.2	116.9

Abbreviations: gCO_2/kJ-Grams of carbon dioxide per kilojoule.

Low-carbon transportation is a green transportation development mode characterized by high energy efficiency, low energy consumption, low pollution, low emission, or even zero-emission [26]. In essence, it is an energy revolution, shifting from fossil energy to green electricity as far as possible. Therefore, the core of developing low-carbon transportation is to improve energy efficiency, optimize energy use structure, and improve service level. To achieve low-carbon and green development of the whole cycle industrial chain in the field of transportation.

4. Model of New Energy Travel Mode Selection in Metropolitan Area under Low-Carbon Orientation

Public transportation plays a significant role in reducing carbon emissions. Public transportation has the advantages of low energy consumption, low emission, and high transportation efficiency, and is a green transportation mode [27]. The construction of an urban, comprehensive transportation system dominated by public transportation has become the consensus of all countries in the world. Improving the sharing rate of public transportation and reducing the use of private transportation will significantly promote the reduction in carbon emissions and the protection of the metropolitan environment.

Commuting travel within the metropolitan area is different from regular inter-city travel and family visits. In the context of increasing travel distance and travel time in metropolitan areas, as the choice of each traveler is an individual behavior, commuters have the problem of choosing different travel tools during rush hour.

Subway, new energy buses, and fuel private cars are the three most common ways for commuters to use. Subway mainly refers to the rail transit built in the city with fast, large volume and electric traction. Compared with the ground bus, the subway has stronger transportation capacity and has the unique advantages of punctuality, fast speed, and saving the land use area of the road surface. Bus mainly refers to the conventional ground bus, with the characteristics of large passenger volume, low fare, low per capita energy consumption, and economic and environmental protection travel mode. The travel time of new energy buses and fuel private cars is uncertain to some extent. However, according to the transportation policy in China, new energy buses can use bus lanes, which will increase the driving speed to a certain extent. Additionally, petrol private cars are not allowed. Therefore, starting from the cumulative prospect theory and expected utility theory, this part fully considers the simulation scenario of commuters' travel behavior, assumes departure time, congestion probability, and possible commuting time consumption, builds a travel mode selection model, calculates the cumulative prospect value and perceived travel cost, and explores the direction of guiding travel behavior by comparing the difference of optimal results under different theoretical frameworks [28].

4.1. *Theoretical Model of Cumulative Prospect Theory*

Prospect theory (PT) introduces psychology into behavioral science for analysis and is developed from expected value theory and expected utility theory by psychology professors Kahneman and Tversky [29]. When observing the behavior of decision-makers in travel behavior, the important feature of prospect theory is that it mainly focuses on the result that travelers may face, that is, the psychological feeling when they gain or lose. According to the prospect theory, under different risk prediction conditions, when people face gains and losses, they will have different feelings based on different reference conditions. Additionally, believes that human behavior tendencies can be predicted [30].

The theory has been widely used to study attitudes toward gain and loss in decision-making. The main content of the theory is as follows: In the premise of failing to make accurate risk judgment, individual behavior decision is determined using the difference between the result and the prior assumption. The decision is composed of the value function and decision power function. It assumes that the uncertain decision process can be divided into two stages: editing and evaluation. Decision makers divide value into gains and losses based on reference points. Changes in gains and losses will change people's subjective feelings about value and thus affect and change people's preferences. In the evaluation stage, the utility function in the expected utility theory is replaced by the value function, the probability of the expected utility function is replaced by the decision weight of the weight function, and the decision is made based on the change of value rather than the current value.

4.2. Commuter Travel Mode Selection Model

According to the idea of cumulative prospect theory, when commuters are faced with a commuting mode choice, they will make decisions according to the following steps: (1) When there is uncertainty in the travel scene and environment, the perceived cost of commuters for each commuting mode is calculated; (2) Aggregate the perceived cost of each commuting mode; (3) Based on previous travel experience, set a travel reference point, which should be as consistent as possible with daily life; (4) On the premise of the above reasonable travel reference points, the perceived travel cost of each travel mode is reasonably judged; (5) To find out whether a travel mode is a benefit or a loss to an individual, and calculate its value; (6) Calculate the cumulative prospect value of each travel mode, that is, accumulate the prospect value and conduct subjective evaluation; (7) After judging and comparing the commuting modes between residence and work place, choose the mode with the maximum cumulative prospect value to commute, and finally complete the decision-making process.

4.2.1. Edit Stage

① How to select the decision reference point has always been the core parameter in prospect theory, which measures the psychological expectations of decision-makers [31]. In the process of travel decision-making, commuters will judge the gains and losses of behavior with certain measurement standards and evaluate the "loss" and "profit" feelings of gains and losses, respectively. Generally, in order to arrive at the destination on time, travelers need to reserve travel time before traveling. The reserved travel time is determined by the travel time between ODs, travel cost, road network status (number of alternative routes), etc., which can be used as a reference point for path selection decisions. Commuters use this reference point to judge whether they arrive early or late, as well as gain and loss.

In this study, commuting time and cost are selected as the reference points for commuters to make decisions. Generally, travelers will determine the attributes of alternative routes based on their own travel purposes and travel needs, on the basis of the effect judgment of the last trip, combined with experience summary, assuming a decision-making reference point and integrating the commuting time and cost. The mathematical formula can be expressed as Equation (1), M_K is the attribute of K alternative path, N is the set of all paths between OD, and N_K is the set of road sections included in path K, ε_α is the road flow, $M_\alpha(\cdot)$ is time function, $\beta_\alpha(\cdot)$ is the cost function, and $P_1 P_2$ is individual preferences, P_1 refers to the coordination of commuters' travel time and cost based on their choice of path; P_2 is a time value parameter, which refers to the degree to which commuters are willing to invest time or money for this travel:

$$M_K = \sum_{\alpha \in N_K} M_\alpha(\varepsilon_\alpha) = \sum_{\alpha \in N_K} [P_1 P_2 M_\alpha(\varepsilon_\alpha) + (1 - P_1)\beta_\alpha(\varepsilon_\alpha)], k \in N \quad (2)$$

Compared with the travel time, if the commuter chooses a certain mode of transportation, the travel cost will be determined accordingly. However, different commuters have different conditions and needs. Therefore, different travel cost reference points will be assumed to select corresponding transportation modes. This paper mainly analyzes the travel choice of public transport and fuel private cars. Based on this, this paper proposes the following hypothesis: office workers have three travel modes: new energy buses, subway, and fueled private cars.

$$CPV = \sum_{i=1}^{n} \pi(p_i) \cdot v(\Delta x_i) \quad (3)$$

Among them, CPV represents the foreground value, $\pi(p_i)$ is the probability weight function of the ith state occurrence, and $v(\Delta x_i)$ is the value function.

② The basic feature of the value function is that a normal person with limited rationality has a risk-averse attitude towards gains or gains and a risk-preference attitude towards losses:

$$v[E(X)] > E[v(X)], E[v(-X)] > v[E(X)], X > 0 \tag{4}$$

The value function describes the psychological utility of loss value and returns value to decision-makers. The value function is described as an S-shaped curve specifically: it is a concave function in the income field and a convex function in the loss field; that is, with the increase in loss value and income value, the marginal utility decreases. This phenomenon is summarized as "decreasing sensitivity". The inflection point of the S-shaped curve, that is, the reference point of decision-making, means that what plays a role in the decision of the decision-maker is not the absolute value of losses and gains but the relative change value relative to the reference point. This feature is summarized as "reference point dependence". The value function is steeper in the loss field than in the income field, which shows that the psychological utility of equal loss is greater than that of income; that is, the decision-maker is more sensitive to loss, which is defined as "loss aversion" [32]. The formula of the value function is shown in Formula (5).

$$v(\Delta x_i) = \begin{cases} \Delta x_i^{\alpha}, \Delta x_i \geq 0 \\ -\lambda(-\Delta x_i)^{\beta}, \Delta x_i < 0 \end{cases} \tag{5}$$

where parameter λ It means that the loss has more influence on the decision-maker than the gain, $\lambda > 1$. Parameters α Additionally, parameters β. It represents the slope of the value curve when facing gains and losses, also known as the risk sensitivity coefficient ($0 < \alpha \leq 1$, $0 < \beta \leq 1$). The recommended parameter values proposed by Kahneman and Tversky are $\alpha = \beta = 0.88$, $\lambda = 2.25$. The specific parameter values are shown in Table 3 [32,33]. X_0 is the decision reference point, Δx is the value of x deviating from the reference point.

Table 3. The value function with the diagram.

Item	Value
α	0.88
β	0.88
λ	2.25
γ	0.61
σ	0.69

③ The weight function describes the decision-maker's subjective perception of probability, which is a probability monotonic increasing function. The formula expression of the decision weight function:

$$H^{+}(p_i) = \frac{p_i^{\gamma}}{\left[p_i^{\gamma} + (1-p_i)^{\gamma}\right]^{\frac{1}{\gamma}}} \tag{6}$$

$$H^{-}(p_i) = \frac{p_i^{\sigma}}{\left[p_i^{\sigma} + (1-p_i)^{\sigma}\right]^{\frac{1}{\sigma}}} \tag{7}$$

4.2.2. Evaluation Stage

The cumulative prospect value is obtained by calculating the cumulative probability of a certain travel mode, taking into account its value function, and the sum of the two products is the cumulative prospect value of the travel mode. The cumulative prospect value of a certain travel mode is as follows:

$$CPV = CPV^{+} + CPV^{-} \tag{8}$$

$$\pi^{+}(p_i) = H(p_i + \cdots + p_n) - H(p_{i+1} + \cdots + p_n); 0 \leq i \leq n-1 \tag{9}$$

$$\pi^{-}(p_i) = H(p_{-mi} + \cdots + p_i) - H(p_{-m} + \cdots + p_{i-1}); 1-m \leq i \leq 0 \tag{10}$$

4.3. Generalized Perceived Travel Cost Function

Assuming that all commuters in the metropolitan area are bounded rational, the cost experienced in the whole travel process is composed of travel time cost and delay cost caused by early arrival and late arrival. The definition of commuter travel cost function is:

$$Total\ Cost_{\mu} = C_{Early} + C_{Late} + C_{Trip} + M_{\mu} \tag{11}$$

Suppose $T_{Departure}$ is the departure time of office workers and $T_{Arrival}$ is the arrival time of office workers. $T_{Arrival} = T_{Departure} + T_{Transit}$, T_{Work} is the working hour specified by the work unit, $E_{ArrivelTime} = T_{Work} - T_{Arrival}$ is the time when the office worker arrives at the work unit early, $L_{ArrivalTime} = T_{Arrival} - T_{Work}$ is the time when the office worker arrives at the work unit late. C_{Trip} is the travel time cost of office workers, $C_{Trip} = \phi \times T_{ActualTransit}$, ϕ refers to the value of commuting travel time for different travel modes, and $T_{ActualTransit}$ refers to the actual duration of commuting for office workers. δ_{Early} indicates the unit time value of early arrival of office workers, δ_{Late} indicates the unit time value of late arrival of office workers. M_{μ} is the transportation cost to be paid for choosing different transportation modes. $1-\rho$ is the additional cost factor of late arrival, ρ is the 0–1 variable, which satisfies the following relationship:

$$\rho = \begin{cases} 0, L_{ArrivalTime} \geq 0 \\ 1, E_{ArrivalTime} > 0 \end{cases} \tag{12}$$

Based on this, the generalized travel cost function can be transformed into:

$$Total\ Cost_{\mu} = C_{Early} + C_{Late} + C_{Trip} + M_{\mu} = \rho \times \delta_{Early}(T_{Work} - T_{Arrival}) + (1-\rho)\delta_{Late}(T_{Work} - T_{Arrival}) + \phi T_{ActualTransit} + M_{\mu} \tag{13}$$

When commuters feel profitable:

$$\Delta Total\ Cost_{\mu} = Total\ Cost_{\mu} - Total\ Cost_{\mu 0} > 0 \tag{14}$$

When commuters feel the loss:

$$\Delta Total\ Cost_{\mu} = Total\ Cost_{\mu} - Total\ Cost_{\mu 0} \leq 0 \tag{15}$$

Based on this, under the cumulative prospect theory, it is assumed that the budgeted travel cost at the reference point of office workers' travel is $Total\ Cost_{\mu 0}$.

4.4. Spatial Structure of Xi'an Metropolitan Area and Data Sources

Spatial Structure of Xi'an Metropolitan Area

On 21 March 2022, the National Development and Reform Commission of China approved the Development Plan of the metropolitan area, which is the fifth metropolitan area plan after the planning of Nanjing, Fuzhou, Chengdu, and Changchun metropolitan area and the only one in northwest China at present. Xi'an metropolitan area is located at the intersection of the horizontal axis of the land bridge passage and the vertical axis of the Bao-kun Passage in China's "two horizontal and three vertical" urbanization strategic pattern. It is the core area of the urban agglomeration of Guanzhong Plain, one of the regions with the best development conditions and the strongest economic and population carrying capacity in the western region, and plays an important role in the overall construction of a modern socialist country and the construction of a new development pattern. The spatial structure evolution diagram of the Xi'an metropolitan area is shown in Figure 3. Xi'an metropolitan area is the latest emerging metropolitan area in China, so it is innovative to study the characteristics and influencing factors of new energy commuting behavior in this area.

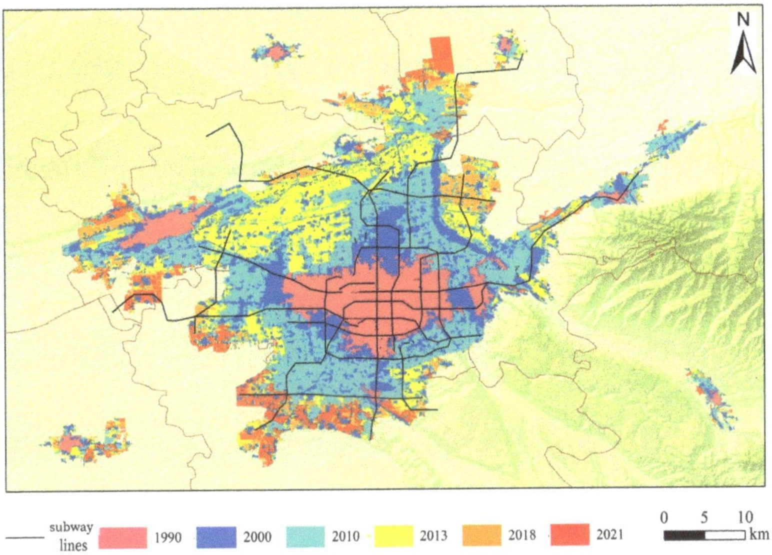

Figure 3. Schematic Diagram of Spatial Expansion of Xi'an metropolitan area.

4.5. Data Sources

In order to comprehensively present the commuting process of commuters in urban areas and link the travel activities of office workers from home to work, this article aims to study the travel behavior of office workers in the context of urban areas. Considering the impact of the epidemic and the limitations of the research scope, an online survey method was adopted for the questionnaire survey. Due to the potential bias or limitations of online surveys, we have adopted two methods in our survey. We have chosen two methods for distributing the online questionnaire. One way is for us to choose locations such as subways and bus stops and directly invite respondents to enter the questionnaire link on-site to fill out the questionnaire. Another method is to select six enterprises distributed in different regions within the Xi'an metropolitan area and entrust their human resources management department to distribute online questionnaires within the enterprises for investigation. The survey was conducted from March 2022 to September 2022, covering the Xi'an metropolitan area. The specific questionnaire design mainly includes understanding the basic information of the respondents, their family economic status, their choice of transportation mode during commuting, as well as the layout of public transportation facilities and personal travel preferences between work and residence. In this survey, the specific investigation content is as follows:

① A survey of basic information about commuter families. This mainly includes the area and street where the household resides, the number of households, the total annual income of the household, and whether the household owns a private car.

② Personal situation survey of commuters. It mainly includes statistics on gender, age, occupation, marital status, registered residence, nature of housing, whether to have a driver's license and length of service.

③ Investigation of personal travel behavior information. This mainly includes the work address of office workers, departure time for commuting, transportation used for commuting, one-way commuting distance and time, one-way commuting fees, and the number of one-way commuting transfers. In actual investigation work, based on the complete process of commuting for a day, the surveyed personnel are required to fill out the entire process from home to work, including the specific location of the stopover location and the means of transportation to be transferred.

5. Result

Then, according to the characteristics of each means of transportation, the commuting time and different possible probabilities brought by the three modes are assumed to compare the selectivity of new energy buses, subways, and fueled private cars in different scenarios.

Mode 1—New energy bus: there is a 70% probability of congestion, travel time is 60 min, there is a 30% probability of no congestion, travel time is 40 min, and the ticket price is 2 yuan;

Mode 2—Subway: The total travel time is fixed at 30 min, and the fare is 4 yuan;

Mode 3—Fuel private cars: the probability of congestion is 60%, the travel time is 45 min, there is a 40% probability of no congestion, the travel time is 35 min, and the cost is 20 yuan;

By setting a scenario, considering the expected possibility of commuters' work time and departure time, the cumulative prospect value of the above method is calculated according to the constraint of the reserved time:

Scenario 1: The commuter's work time is 8:00, departure time is 7:20, and needs to arrive at work within 40 min;

Scenario 2: The commuter's work time is 8:00, departure time is 7:10, and needs to arrive at work within 50 min;

Scenario 3: Commuters start work at 8:00, depart at 7:00, and need to arrive at work within 60 min.

By setting the scenario, considering the expected possibility of the working time and departure time of office workers, according to the constraints of the reserved time, calculate the cumulative prospect value of the above methods, randomly distribute 1000 questionnaires, and recover 860 valid questionnaires, with an effective recovery rate of 86%. According to the minimum living security standard of 740 yuan per person per month for urban residents in Xi'an from 1 October 2020 and the maximum size of conventional families as the standard, families with annual income less than 50,000 yuan are defined as low-income families and other families are classified as non-low-income families.

According to the survey data, 518 men and 342 women commuted among 860 people, accounting for 67.21% and 39.77% of the total, respectively. In terms of age distribution, there are 65 people under the age of 20, 269 people aged 20–29, 314 people aged 20–39, 177 people aged 40–49, and 35 people aged 50–59. In terms of occupational attributes, civil servants account for 14.65%; public institutions staff account for 19.42%; state-owned enterprises 25.93%; private enterprise staff 29.77%; and foreign enterprises 10.23%. From the distribution of seniority, new employees within 2 years accounted for 6.63%, those within 2–5 years accounted for 21.87%, those within 5–10 years accounted for 30.81%, those between 10–20 years accounted for 32.68%, and those over 20 years accounted for 8.02%. The descriptive statistics of the personal survey results of commuters are shown in Table 4.

Tables 5–10 show the perceived costs and cumulative prospect values of traveler decision-making under three different scenarios calculated through the model.

Based on specific data, the following conclusions can be drawn: (1) From Table 5, it is found that under the expected utility theory, commuters believe that the subway is the optimal mode of transportation; (2) From Table 6, it can be seen that commuters believe that new energy bus has the highest returns. (3) From Table 7, it is found that under the expected utility theory, commuters consider the subway to be the optimal mode of transportation (4) From Table 8, it can be seen that commuters believe that the subway has the highest revenue. (5) From Table 9, it is found that under the expected utility theory, commuters consider the subway to be the optimal mode of transportation. (6) From Table 10, it can be seen that commuters believe that new energy buses have the highest returns.

Table 4. Descriptive Statistics of Personal Survey Results for Commuters.

Item	Description	Number (N = 860)	Percentage (%)
Age	Under 20 year	65	7.90
	20–29 years old	269	31.28
	30–39 years old	314	36.51
	40–49 years old	177	20.58
	50–59 years old	35	4.07
Gender	Male	518	67.21
	Female	342	39.77
Occupation	Civil servant	126	14.65
	Public institutions staff	167	19.42
	Staff of state-owned enterprise	223	25.93
	Private enterprise staff	256	29.77
	Staff of foreign enterprise	88	10.23
Working experience	0–2 years	57	6.63
	2–5 years	188	21.87
	5–10 years	265	30.81
	10–20 years	281	32.68
	20 years–∞	69	8.02
Does the family own a car	Yes	475	55.23
	No	385	44.77
Annual household income	0–50,000 yuan	70	8.13
	50,000–100,000 yuan	158	18.37
	100,000–150,000 yuan	193	22.44
	150,000–200,000 yuan	200	23.25
	200,000–300,000 yuan	138	16.05
	300,000 yuan–∞	101	11.74

Table 5. Expected Travel Costs for Different Transportation Modes under Scenario 1.

	Fuel Private Cars	Subway	New Energy Bus
Expected travel time	60 min, 70% 40 min, 30%	30 min	45 min, 60% 35 min, 40%
Perceived travel costs	92.7, 70% 31.6, 30%	56.4	114.23, 60% 53.756, 40%
Expected value of travel cost	63.51	56.4	87.311

Table 6. The cumulative prospect values of different modes of transportation in Scenario 1.

	Fuel Private Cars	Subway	New Energy Bus
Travel cost reference point	24.08	25.55	62.3
Travel cost function value	−84.23, 70% 0, 30%	−40.72	−66.34, 60% 0.85, 40%
CPV	−43.32	−47.43	−35.28

Table 7. Expected travel costs for different modes of transportation in Scenario 2.

	Fuel Private Cars	Subway	New Energy Bus
Expected travel time	60 min, 70% 40 min, 30%	30 min	45 min, 60% 35 min, 40%
Perceived travel costs	70.63, 70% 53.6, 30%	27.5	88.316, 60% 70.481, 40%
Expected value of travel cost	67.4	27.5	72.606

Table 8. The cumulative prospect values of different modes of transportation in Scenario 2.

	Fuel Private Cars	Subway	New Energy Bus
Travel cost reference point	25.3	32.25	81.4
Travel cost function value	−51.46, 70% −22.37, 30%	−1.66	−9.16, 60% 1.47, 40%
CPV	−24.21	−21.34	−6.2

Table 9. Expected travel costs for different modes of transportation in Scenario 3.

	Fuel Private Cars	Subway	New Energy Bus
Expected travel time	60 min, 70% 40 min, 30%	30 min	45 min, 60% 35 min, 40%
Perceived travel costs	62.4, 70% 55.42, 30%	51.23	83.27, 60% 77.31, 40%
Expected value of travel cost	56.03	53.141	79.022

Table 10. The cumulative prospect values of different modes of transportation in Scenario 3.

	Fuel Private Cars	Subway	New Energy Bus
Travel cost reference point	47.3	37.21	76.54
Travel cost function value	−30.4, 70% −26.06, 30%	−12.4	0.636, 60% 0.709, 40%
CPV	−28.15	−13.17	0.68

By integrating the results of the above three scenarios, it can be concluded that under different travel constraints, commuters use travel costs as a reference point, and the cumulative prospect values obtained are shown in Figure 4. The results of the above analysis show that (1) commuters will be affected by reference points in the process of travel behavior selection, which is consistent with the theoretical content of cumulative prospect theory. This is consistent with the results of the study, mainly because commuters may conduct empirical evaluations before choosing their mode of transportation, which proves the importance of further understanding commuters' judgments of the travel environment before traveling [34]. (2) In the simulation, it is found that it is effective to take generalized travel costs as the reference point, and commuters will make rational judgments according to the actual situation. Under the premise of sufficient reservation time and ensuring that there will be no late for work, commuters will prefer to choose the more secure means of transportation with low congestion probability when faced with benefits. This is different from previous studies, which have suggested that commuters are irrational when choosing transportation, and their personal psychological preferences are difficult to change [35]. (3) If commuters reserve a short time and find that they are likely to be late through experience judgment, they will turn into adventurers and form a "gambler's psychology" when faced with losses and are more likely to choose transportation with greater flexibility and a probability of arriving at work in a short time, such as fuel private cars [36]. (4) Different scenarios and assumptions will cause commuters to make different travel decisions. Commuters tend to evaluate different choices through the judgment criteria of utility maximization, and there are differences between traveler choice results and expectation theory.

In summary, it can be found that under the low-carbon orientation, commuters in the Xi'an metropolitan area generally believe that the subway has obvious advantages under different assumptions, which is in line with the current development direction of optimizing transportation energy in the metropolitan area; In addition, in the scenario, due to the support of national energy policies in China, new energy buses have the qualification to enjoy dedicated bus lanes in infrastructure construction. Therefore, they can avoid road congestion during commuting, which is not available in gasoline private cars. Meanwhile, in scenario three, new energy buses and subways, as new energy public transportation modes, have significant benefits.

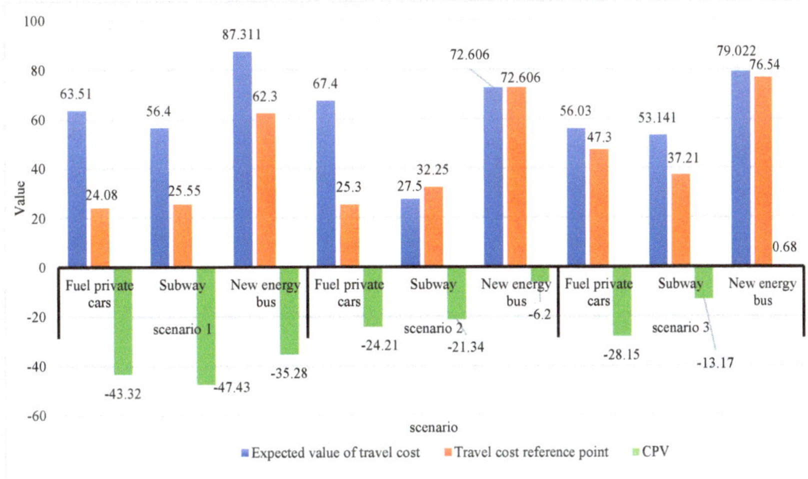

Figure 4. The Cumulative Foreground Model Results in Three Scenarios.

6. Conclusions and Suggestion

This article takes the Xi'an metropolitan area as an example to elaborate on the integrated development path of the new energy + transportation industry chain in the metropolitan area and clarifies the energy transformation model for the integrated development of low-carbon transportation energy. From the perspective of green and low-carbon, the cumulative prospect model was used to verify the internal mechanism that affects commuters in metropolitan areas to choose new energy commuting modes. The research results indicate that new energy transportation modes play an important role in a low-carbon economy, and there are significant differences in the cumulative prospect values of subways, new energy buses, and gasoline private cars under different scenarios and assumptions.

Therefore, we believe that (1) low-carbon-oriented commuting in urban areas is easily influenced by the characteristics of transportation modes. Subways and new energy buses have obvious advantages in energy optimization for commuting, which has become one of the directions for the future development of low-carbon transportation in urban areas. (2) Commuters face a significant threat to the proportion of private fuel cars traveling due to the significant advantages of new energy public transportation in terms of commuting time and cost when facing the choice of transportation tools. In order to expand the proportion of public transportation, such as new energy buses and subways in daily commuting, we need to improve the construction of public transportation infrastructure and increase the burden of using fuel-powered private cars. (3) Accelerating the proportion of new energy in public transportation is the key to reducing carbon emissions from public transportation. New energy public transportation has lower energy consumption and emission levels, which helps promote the application of new energy and low-carbon technologies.

However, it is important to also acknowledge the limitations of this study. In terms of case selection and data investigation, we have focused on China. Taking the newly approved urban agglomeration in western China as an example, although it has some innovation in the research area, we have not taken into account other mature urban agglomerations in China. At the same time, relying on big data methods, the number of data samples can be increased to compensate for the subjective bias in data in order to improve data reliability, which will help to study this topic better.

Based on these findings, we provide the following suggestions for improving the proportion of new energy commuting in the Xi'an metropolitan area:

6.1. Build a Complete Network of Ground Bus Charging Facilities

For large cities in China, the first step should be to take measures to strictly control the growth and use of private cars. Measures such as traffic restrictions, license plate restrictions, differential parking fees, congestion fees, and staggered commuting should continue to be implemented to avoid traffic congestion. A new energy vehicle charging pile is one of the key areas of "new infrastructure", accelerates the construction of the charging facilities network, on the one hand, strengthens the technological innovation of charging facilities, strengthens the digital gene, and promotes the deep integration of traditional charging facilities, ground bus operation network and new technologies such as artificial intelligence, block-chain, and big data [37]. Actively explore the construction of an intelligent network platform from the planning and construction of front-end charging facilities to the deployment of intermediate bus charging needs and then to the management and maintenance of terminal charging facilities. On the other hand, strengthen the innovation of charging operation mode, face the subdivision scenarios of the charging demand of public vehicles and social vehicles at different times, take into account safety, efficiency, and energy saving, create technology applications such as wireless charging of charging piles and customized charging management of vehicles, and form an ecological model of multi-type charging facility investment, diversified charging methods, and diversified profit sharing [38]. Maximize the utilization rate and profitability of charging facilities to match the charging demand and management level.

6.2. Promote the Transformation of the New Energy Travel Structure

The transformation of motor vehicle energy structure is the core of promoting urban transportation emission reduction, and it is also the most potential strategy. To achieve a carbon peak in 2030, first rely on the decarbonization of the energy system and the decarbonization of the energy system depends on the energy storage of new energy vehicles, and the new energy revolution is driven by new energy vehicles [39]. As a representative of new energy vehicles, pure electric vehicles are the integrated products of modern automotive technology, new energy, electronic computer intelligent control, and other high-tech, which do not produce CO_2 during operation and use and have the advantages of environmental protection and pollution-free, high energy efficiency, and low operating costs. The emergence and popularization of pure electric vehicles can not only make the automobile industry get rid of the situation of excessive dependence on gasoline but also reduce carbon emissions, and the emission reduction effect is remarkable [40]. Attention should be paid to the planning and construction of energy supply facilities, improving the power supply and grid capacity of cities, and strengthening the construction of charging facilities.

6.3. Strengthen the Publicity of Low-Carbon Travel and Create a Low-Carbon Travel Cultural Atmosphere

Through the media, the internet, public service advertising, and other advertising media, the government can inform urban residents about the growing problems of traffic congestion, air pollution, and energy consumption and their serious consequences so that citizens can respond to the challenges posed by motorization and make positive contributions to reducing air pollution and traffic congestion [41]. The government should also step up efforts to promote low-carbon living and low-carbon travel to the public, especially the high-income group, and encourage the public to travel using means of transport that minimizes damage to the environment through extensive publicity on energy conservation to reduce carbon emissions and the implementation of low-carbon emission practices.

Author Contributions: Conceptualization, X.D.; methodology, X.D.; software, X.D.; validation, X.D. and T.M.; formal analysis, X.D.; investigation, X.D., T.M. and E.Z.; resources, X.D.; data curation, X.D.; writing—original draft preparation, X.D.; writing—review and editing, X.D.; visualization, X.D.; supervision, T.M. and E.Z. All authors have read and agreed to the published version of the manuscript.

Funding: This research received no external funding.

Informed Consent Statement: The survey questionnaire designed in this study did not involve the privacy of the respondents during the distribution process and was explained to the respondents in the first line of the questionnaire. Therefore, when respondents agree to answer the questionnaire, they indicate informed consent. The questionnaire data is anonymous and fully applicable to academic research.

Data Availability Statement: Data are contained within the article.

Acknowledgments: We would like to thank the editors and anonymous reviewers for their constructive suggestions and comments, which helped to improve this paper's quality.

Conflicts of Interest: The authors declare no conflict of interest.

References

1. Glasby, G.P. Sustainable development: The need for a new paradigm. *Environ. Dev. Sustain.* **2002**, *4*, 333–345. [CrossRef]
2. Zhang, H.; Ding, T.; Sun, Y.; Huang, Y.; He, Y.; Huang, C.; Li, F.; Xue, C.; Sun, X. How does load-side re-electrification help carbon neutrality in energy systems: Cost competitiveness analysis and life-cycle deduction. *Renew. Sustain. Energy Rev.* **2023**, *187*, 113745. [CrossRef]
3. Li, J.; Jiao, L.; Li, F.; Lu, X.; Hou, J.; Li, R.; Cai, D. Spatial disequilibrium and influencing factors of carbon emission intensity of construction land in China. *J. Clean. Prod.* **2023**, *396*, 136464. [CrossRef]
4. Tseng, Y.; Yue, W.L.; Taylor, M.A.P. The role of transportation in logistics chain. *East. Asia Soc. Transp. Stud.* **2005**, *5*, 1657–1672.
5. Fernández-Dacosta, C.; Shen, L.; Schakel, W.; Ramirez, A.; Kramer, G.J. Potential and challenges of low-carbon energy options: Comparative assessment of alternative fuels for the transport sector. *Appl. Energy* **2019**, *236*, 590–606. [CrossRef]
6. Holz-Rau, C.; Scheiner, J. Land-use and transport planning—A field of complex cause-impact relationships. Thoughts on transport growth, greenhouse gas emissions and the built environment. *Transp. Policy* **2019**, *74*, 127–137. [CrossRef]
7. Wyatt, D.W.; Li, H.; Tate, J.E. The impact of road grade on carbon dioxide (CO_2) emission of a passenger vehicle in real-world driving. *Transp. Res. Part D Transp. Environ.* **2014**, *32*, 160–170. [CrossRef]
8. Dällenbach, N. Low-carbon travel mode choices: The role of time perceptions and familiarity. *Transp. Res. Part D Transp. Environ.* **2020**, *86*, 102378. [CrossRef]
9. Fletcher, J.; Longnecker, N.; Higham, J. Envisioning future travel: Moving from high to low carbon systems. *Futures* **2019**, *109*, 63–72. [CrossRef]
10. Shie, A.-J.; Dai, Y.-Y.; Shen, M.-X.; Tian, L.; Yang, M.; Luo, W.-W.; Wu, Y.J.; Su, Z.-H. Diamond model of green commitment and low-carbon travel motivation, constraint, and intention. *Int. J. Environ. Res. Public Health* **2022**, *19*, 8454. [CrossRef]
11. Liao, C.; Huang, Y.; Zheng, Z.; Xu, Y. Investigating the factors influencing urban residents' low-carbon travel intention: A comprehensive analysis based on the TPB model. *Transp. Res. Interdiscip. Perspect.* **2023**, *22*, 100948. [CrossRef]
12. Moriarty, P. Making urban travel sustainable: Travel reductions are needed. *Clean. Prod. Lett.* **2022**, 3, 100010. [CrossRef]
13. Næss, P. Compact urban development in Norway: Spatial changes, underlying policies and travel impacts. In *Advances in Transport Policy and Planning*; Academic Press: Cambridge, MA, USA, 2022; Volume 9, pp. 95–133.
14. Abdul-Manan, A.F.N. Uncertainty and differences in GHG emissions between electric and conventional gasoline vehicles with implications for transport policy making. *Energy Policy* **2015**, *87*, 1–7. [CrossRef]
15. Broin, E.Ó.; Guivarch, C. Transport infrastructure costs in low-carbon pathways. *Transp. Res. Part D Transp. Environ.* **2017**, *55*, 389–403. [CrossRef]
16. Chavhan, S.; Venkataram, P. Prediction based traffic management in a metropolitan area. *J. Traffic Transp. Eng. Engl. Ed.* **2020**, *7*, 447–466. [CrossRef]
17. Walsh, M.P. Motor vehicle pollution and fuel consumption in China: The long-term challenges. *Energy Sustain. Dev.* **2003**, *7*, 28–39. [CrossRef]
18. Chang, X.; Ma, T.; Wu, R. Impact of urban development on residents' public transportation travel energy consumption in China: An analysis of hydrogen fuel cell vehicles alternatives. *Int. J. Hydrogen Energy* **2019**, *44*, 16015–16027. [CrossRef]
19. O'Malley, M.J.; Anwar, M.B.; Heinen, S.; Kober, T.; McCalley, J.; McPherson, M.; McPherson, M.; Muratori, M.; Orths, A.; Ruth, M.; et al. Multicarrier energy systems: Share our energy future. *Proc. IEEE* **2020**, *108*, 1437–1456. [CrossRef]
20. Shi, B.; Wu, L.; Kang, R. Clean Development, energy substitution, and carbon emissions: Evidence from clean development mechanism (CDM) project implementation in China. *Sustainability* **2021**, *13*, 860. [CrossRef]

21. Hadjipaschalis, I.; Poullikkas, A.; Efthimiou, V. Overview of current and future energy storage technologies for electric power applications. *Renew. Sustain. Energy Rev.* **2009**, *13*, 1513–1522. [CrossRef]
22. Yang, C.; Zha, M.; Wang, W.; Liu, K.; Xiang, C. Efficient energy management strategy for hybrid electric vehicles/plug-in hybrid electric vehicles: Review and recent advances under intelligent transportation system. *IET Intell. Transp. Syst.* **2020**, *14*, 702–711. [CrossRef]
23. Zeng, Y.; Tan, X.; Gu, B.; Wang, Y.; Xu, B. Greenhouse gas emissions of motor vehicles in Chinese cities and the implication for China's mitigation targets. *Appl. Energy* **2016**, *184*, 1016–1025. [CrossRef]
24. Elgowainy, A.; Rousseau, A.; Wang, M.; Ruth, M.; Andress, D.; Ward, J.; Joseck, F.; Nguyen, T.; Das, S. Cost of ownership and well-to-wheels carbon emissions/oil use of alternative fuels and advanced light-duty vehicle technologies. *Energy Sustain. Dev.* **2013**, *17*, 626–641. [CrossRef]
25. Wang, Z.; Chen, F.; Fujiyama, T. Carbon emission from urban passenger transportation in Beijing. *Transp. Res. Part D Transp. Environ.* **2015**, *41*, 217–227. [CrossRef]
26. Lee, C.T.; Hashim, H.; Ho, C.S.; Van Fan, Y.; Klemeš, J.J. Sustaining the low-carbon emission development in Asia and beyond: Sustainable energy, water, transportation and low-carbon emission technology. *J. Clean. Prod.* **2017**, *146*, 1–13. [CrossRef]
27. Jiang, Y.; Zhou, Z.; Liu, C. The impact of public transportation on carbon emissions: A panel quantile analysis based on Chinese provincial data. *Environ. Sci. Pollut. Res.* **2019**, *26*, 4000–4012. [CrossRef] [PubMed]
28. An, S.; Hu, X.; Wang, J. A cumulative prospect theory approach to car owner mode choice behaviour prediction. *Transport* **2014**, *29*, 386–394. [CrossRef]
29. Tversky, A. A critique of expected utility theory: Descriptive and normative considerations. *Erkenntnis* **1975**, *9*, 163–173.
30. McDermott, R.; Fowler, J.H.; Smirnov, O. On the evolutionary origin of prospect theory preferences. *J. Politics* **2008**, *70*, 335–350. [CrossRef]
31. Pachur, T.; Schulte-Mecklenbeck, M.; Murphy, R.O.; Hertwig, R. Prospect theory reflects selective allocation of attention. *J. Exp. Psychol. Gen.* **2018**, *147*, 147–169. [CrossRef]
32. Kahneman, D.; Tversky, A. Choices, values, and frames. *Am. Psychol.* **1984**, *39*, 341. [CrossRef]
33. Avineri, E.; Prashker, J.N. Violations of expected utility theory in route-choice stated preferences: Certainty effect and inflation of small probabilities. *Transp. Res. Rec. J. Transp. Res. Board* **2004**, *1894*, 222–229. [CrossRef]
34. Geng, K.; Wang, Y.; Cherchi, E.; Guarda, P. Commuter departure time choice behavior under congestion charge: Analysis based on cumulative prospect theory. *Transp. Res. Part A Policy Pract.* **2023**, *168*, 103564. [CrossRef]
35. Gao, K.; Yang, Y.; Sun, L.; Qu, X. Revealing psychological inertia in mode shift behavior and its quantitative influences on commuting trips. *Transp. Res. Part F Traffic Psychol. Behav.* **2020**, *71*, 272–287. [CrossRef]
36. Frei, C.; Hyland, M.; Mahmassani, H.S. Flexing service schedules: Assessing the potential for demand-adaptive hybrid transit via a stated preference approach. *Transp. Res. Part C Emerg. Technol.* **2017**, *76*, 71–89. [CrossRef]
37. Wang, M.; Wu, F.; Chen, J. Comprehensive analyses of the spatio-temporal variation of new-energy vehicle charging piles in China: A complex network approach. *Front. Phys.* **2021**, *9*, 755932. [CrossRef]
38. Wu, Z.; Wang, J.; Zhong, H.; Gao, F.; Pu, T.; Tan, C.W.; Chen, X.; Li, G.; Zhao, H.; Zhou, M.; et al. Sharing economy in local energy markets. *J. Mod. Power Syst. Clean Energy* **2022**, *11*, 714–726. [CrossRef]
39. Jafari, M.; Botterud, A.; Sakti, A. Decarbonizing power systems: A critical review of the role of energy storage. *Renew. Sustain. Energy Rev.* **2022**, *158*, 112077. [CrossRef]
40. Gillingham, K.; Stock, J.H. The cost of reducing greenhouse gas emissions. *J. Econ. Perspect.* **2018**, *32*, 53–72. [CrossRef]
41. Rajé, F.; Tight, M.; Pope, F.D. Traffic pollution: A search for solutions for a city like Nairobi. *Cities* **2018**, *82*, 100–107. [CrossRef]

www.ingramcontent.com/pod-product-compliance
Lightning Source LLC
Chambersburg PA
CBHW061548210926
51850CB00009B/2037
9783725811991

MDPI
St. Alban-Anlage 66
4052 Basel
Switzerland
www.mdpi.com

Energies Editorial Office
E-mail: energies@mdpi.com
www.mdpi.com/journal/energies